What Primary Teachers Should Know About

MATHS

Aileen Duncan

Hodder & Stoughton

LONDON SYDNEY AUCKLAND

British Library Cataloguing in Publication Data

Duncan, Aileen
    What Primary Teachers Should Know About
    Maths
    I. Title
    372.7

ISBN 0 340 55954 3

First published 1993

© 1992 Aileen Duncan

Typeset by Wearset, Boldon, Tyne and Wear
Printed in Great Britain for the educational publishing
division of Hodder & Stoughton Ltd, Mill Road, Dunton
Green, Sevenoaks, Kent by Thomson Litho Ltd

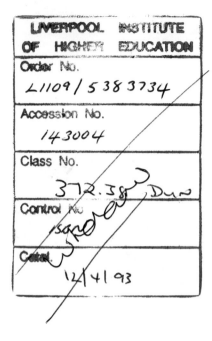

Contents

Introduction

The National Curriculum in England, the Northern Ireland Curriculum and the National Guidelines for 5–14 in Scotland make demands on teachers. School staff and students in training are required to flesh out 'programmes of study' for mathematics to create school policy documents, class forward plans and daily learning experiences. Student teachers and many teachers in post seek support to be more confident about these tasks. This reference book has been written to offer such support.

The book has three sections:

Section One   Aspects of mathematics
Section Two   Teaching and learning
Section Three   Example lesson notes

SECTION ONE – ASPECTS OF MATHEMATICS

This section is intended as a reference for a wide range of mathematical content. The format used is 'question and answer'. The questions are based on what the reader is likely to want to know about each aspect. The format allows individual questions to be easily located.

This section provides support by helping readers to:

- check on their present knowledge of mathematics
- extend their knowledge
- increase their vocabulary to include less familiar terms
- relate each aspect of mathematics to some appropriate contexts
- be more confident about the development of a particular aspect
- be aware of some suitable resources
- be aware of learning difficulties which can arise within each aspect of mathematics
- devise and carry out relevant assessment.

SECTION TWO – TEACHING AND LEARNING

Teachers like to have variety in their teaching as they realise this makes learning so much more interesting for them and their pupils. Different approaches to learning and teaching are considered with particular relevance to mathematics.

This section provides support by helping readers to:

- rethink the range of learning and teaching approaches which they use
- realise which approach might be used for the teaching and learning of specific concepts and procedures
- integrate aspects of mathematics to other work in the curriculum
- plan and organise the use of games, mathematical equipment, the calculator and the computer
- use workcards, worksheets and textbooks more effectively.

SECTION THREE – EXAMPLE LESSON NOTES

Example lessons are used to illustrate how a particular aspect of mathematics can be taught using one of the approaches to learning and teaching.

This section provides support by helping readers to:

- select different approaches for familiar aspects of mathematics
- consider ways of presenting aspects which are less familiar
- differentiate learning tasks to meet the different needs of groups of children
- think about the amount of 'new' learning introduced and how the steps are linked together in a sequence of lessons.

SECTION ONE
ASPECTS OF MATHEMATICS

O N E

Introduction

This section is intended as a reference for a wide range of mathematical content which is grouped under headings of:

- Kinds of number
- Using numbers
- Measures
- Shape, position and movement
- Information handling.

The format used is 'question and answer'. The same questions are used for each aspect of mathematics. The questions are based on what the reader is likely to want to to know. Answers are not exhaustive otherwise this book would have been like Topsy, and 'growed and growed'. The main purpose is to supply basic information and practical ideas.

What does ... mean?

In the answers to this question, each aspect of mathematics is explained. This provides a basis for your discussions with your pupils. It is important that mathematics is not the silent subject of the curriculum but the focus of discussion. Encourage the children to express their understanding and opinions so that you can build on these.

What are real life experiences of ... ?

Mathematics often seems confined to the pages of a textbook. Real life examples help to give a meaningful context to the aspect of mathematics. A few are listed. You and the pupils can add others.

What is the key vocabulary and what does each word mean?

A short explanation is given of the main vocabulary associated with each aspect. The pupils should be asked to build wordbanks. They could think of describing words, of naming words and of action words for each aspect of mathematics. Some of these words will be known to them, others will be added as lessons take place. Each word can be written on a flashcard and the set of flashcards used for reading practice and discussion. The flashcards can also be displayed on a poster.

What is the historical background of ... ?

There is no attempt to provide any detailed or chronological background, but rather some anecdotes which the teacher might pass on to pupils. The ancient peoples developed mathematics because there was a need to describe, compare, order, quantify and record. Once this process was underway, the early concepts and skills soon became established as a subject to be investigated and extended. A process which is still happening today. Many famous mathematicians have left a legacy to us. When the children hear about these mathematicians they may be motivated to ask you to find out more for them, or to read some mathematical history for themselves.

What is the value of teaching . . . ?

Children often wonder why they are learning about some aspects of mathematics. Teachers often wonder why they are teaching some aspects of mathematics. A few justifications are offered.

What are possible key steps in development for the learner?

The suggested developments are not exhaustive but are intended to provide some directions.

What are appropriate resources for teaching . . . ?

Only some suggestions are included but they include the usual, and sometimes the unusual.

What are possible contexts through which . . . might be taught?

A context is the setting in which the mathematical aspect is taught. Sometimes this is related to real life, to simulations, and to games. Sometimes the mathematics can arise out of theme, topic or project work. Sometimes the mathematics can be integrated with another subject or subjects. Where there is the opportunity to use a context, the suggestions given might be used or adapted.

How might . . . be assessed?

Assessment of mathematics could involve oral, practical, written and problem-solving responses from the pupils. Some suggestions are given for each category to help you to devise your own assessment items.

What are common difficulties which children encounter and how might these be overcome?

Common learning difficulties are identified and some remedial action suggested.

TWO

About numbers

WHOLE NUMBERS – CARDINALITY, ORDINALITY, COUNTING

What does whole number mean?

Number names and symbols have been devised to communicate the amount of objects which are in any set. These numbers are called *natural numbers*. For example, when we see a set with this many *** we say 'three' and record the symbol or numeral 3.

Number names and numerals were devised to match in a one-to-one correspondence with sets of one member, of two members, and so on to an infinite amount. However, as well as associating a number name and numeral with a set for a specific amount, the numbers (name and numeral) are used for the process of counting and then for operations such as addition and subtraction.

As the system of recording and using numbers became more sophisticated, a new number name and corresponding numeral was devised – zero or 0. This extended set of numbers – all the natural numbers and zero – make up the *whole numbers*.

What are real life examples of whole numbers?

Whole numbers are used in different ways, for example:

- to represent how many are in a set – referred to as the *cardinality* of the set – as in 'the book has 247 pages'
- to identify the position of one within a set – called the *ordinal* aspect – as in 'turn to page 63 in the book'
- to label an item, for example, A4, a specific size of paper, and a 10 bus indicating a specific route the bus takes.

What is the key vocabulary and what does each word mean?

Some vocabulary has already been explained above, for example, cardinality, ordinality, whole and natural numbers. Other words include:

Numeral – the written symbol(s) for a number, used for a single symbol or arrangement of them, for example 3 is a numeral and so is 4567; the word 'number' is often used instead of the more precise 'numeral'.

Digit – the single symbol for a number as in 'write a numeral with three digits'; the word 'figure' has the same meaning.

Counting – number names and numerals can be ordered by placing sets of objects for the relationship 'one less than' as shown here:

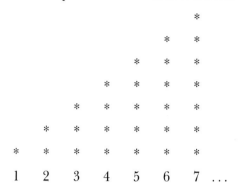

$$1 \quad 2 \quad 3 \quad 4 \quad 5 \quad 6 \quad 7 \ldots$$

Saying the number names in the resulting order of ascending magnitude is called counting. Numbers are used in counting order to find out the cardinality of any set. When objects are counted each number refers to all the objects which have been counted at that time so it is useful practice for young children to physically pick up each object and place it in a pile as they count. This means they see a (sub)set which matches the number they say.

three

Children should realise that they can count by giving a number name to each object in any order but they must give only one number name to each object and every object must be given a number name, for example:

```
*     *     *     *     *      can be counted as
one  two three four  five              or
*     *     *     *     *
       one  two three
four  five                             or
*     *     *     *     *
five four three two  one      and so on . . .
```

Composition of number – the number in a set is not easy to recognise, so the eye breaks the set down into subsets, for example the eye might see a set of six as a set of four and a set of two or two sets of three.

Operations – the mathematical processes of addition, subtraction, multiplication and division.

Calculation – to determine a number or value from information given by means of a mathematical procedure.

Algorithm – a step-by-step procedure by which an operation can be carried out.

What is the historical background of whole numbers?

At one time an amount was recorded by a matching process, for example by collecting stones, by making knots, by carving notches or by marking strokes for each item. Gradually this one-to-one correspondence was replaced by a system of numeration where recorded symbols were introduced. For example, the Egyptians made strokes and other symbols on parchment and the Babylonians made wedge-shaped symbols in clay.

Egyptian   Babylonian

These early systems were mainly used for recording numbers and not for calculations.

What is the value of teaching whole numbers?

The children should be able to:
- interpret symbols used to represent numbers
- say the number names
- read the number names
- write the number names and numerals.

Such skills are essential to enable the children to communicate about amounts and about the position of one member within a set. Although children should know what a number is and be able to use numbers, it is not necessary for them to know terms such as cardinality, ordinality, natural numbers and whole numbers.

What are possible key steps in development for the learner?

The number system is sophisticated and the children learning it are very young, so the steps should be well sequenced and carefully taught. Here is a possible development.

1 Pre-number work
As a preparation for number, children could sort items to form sets. As they label a set with the definition 'red', they are preparing themselves for a set having an attribute such as 'two'. Matching pairs of sets to find which has more, which has as many as, and which has not as many/less/fewer, should enable them to cope more effectively with number comparisons later.

2 Numbers one to five
- Cardinality – matching number names (one, two, three, four, five) to sets and creating sets with one to five members, preferably not in counting order.
- Comparison – comparing sets to find that, for example, four is more than two.
- Composition – recognising subsets in a set, for example, a set of three is made up of a subset of one and a subset of two.
- Counting – putting sets in order to show the relationship 'one less than' and to give the sequence of number names, one to five.
- Learning to recognise and how to write the numerals 1, 2, 3, 4 and 5 and relating each of these to the appropriate set and number name.

3 Numbers six to ten
- Cardinality – matching number names (six, seven, eight, nine, ten) to sets and creating

sets with six to ten members, preferably not in counting order.
- Comparison – comparing sets to find that, for example, eight is more than six.
- Composition – recognising subsets within the sets, for example, a set of six is made up of a set of one and a set of five (being aware of composition helps the child to recognise numbers in this range).
- Counting – putting sets in order to show the relationship 'one less than' and to give the sequence of number names, six to ten.
- Learning to recognise and to write the numerals 6, 7, 8, 9 and 10 and relating each to the appropriate set and number name.

4 Numbers one to ten
- Comparison – comparing sets to find that, for example, seven is more than three.
- Counting – putting sets in order to show the relationship 'one less than ' and to give the sequence of number names, one to ten.
- Reading – being able to say 'one' for both the written one and 1, similarly for the other number names and numerals.
- Ordinality – 'first' and 'last' may be used as a beginning to the use of number to denote position.
- Zero – introduced often through subtraction as the number name to represent 'none'; the numeral is 0.

A further development of whole numbers is found under the section entitled **Whole Numbers – place value.**

What are appropriate resources for teaching whole numbers?

The children could use real objects and pictures or drawings to help them to realise that number is used to find how many objects there are. It is interesting for the children to give a purpose to their number tasks either as part of the daily class routine or within the context of theme work. Unifix blocks are useful to represent sets of different cardinality either as discrete objects or as a continuous tower. Children can then think about, for example, three as three separate ones or as one set of three.

What are possible contexts through which whole numbers might be taught?

Any context gives scope for simple number work. 'Myself' is particularly suitable as the parts of the body can be described in number terms, such as, two eyes, one nose; clothing can also be thought of in number terms, for example, two socks, three buttons on my cardigan; relations can be considered as one aunt, three uncles, two sisters, four grandparents and so on.

How might whole numbers be assessed?

Assess *orally* by asking children to:
- say the number name for selected sets in a photograph
- find which set in given pairs of sets has more
- count on from any number
- state a number which is more/less than a given number.

Assess *practically* by asking children to:
- make a set with as many items as a given set
- create a set for a given number using a rubber stamp, for example, of an animal
- make a set for 'more than 6' using gummed shapes.

Assess *in written form* by asking children to:
- draw a set with more/as many as a given number
- label some drawings of sets with numerals.

Assess through *problem solving* by asking children to:
- choose a number, then make up and tell a story in which the number is often used.

What are common difficulties which children encounter and how might these be overcome?

Counting – some children believe that the single object which they handle when they say a number name represents that number. This makes it very important *not* to use the form of display shown on the left, but to use the one shown on the right.

*	*	*	*	*	**	***	****
1	2	3	4	1	2	3	4

Recording numbers – some children believe each numeral to be made up of the appropriate number of strokes. For example, this is true for 1 but not for other numerals. If they believe 2 has two strokes, they have difficulty in making an acceptable numeral.

$$\bar{} \text{ not } Z$$

It needs to be explained that the numeral is a shorthand way of writing the number name and does not show 'how many'. Another difficulty which may arise when children record numbers is the orientation. They show the numeral facing the opposite direction or lying horizontally.

$$\varepsilon \quad \omega$$

Appoint a child with this difficulty as a 'silly number' watcher with the task of finding numerals which are going backwards or lying on their backs. This can help the child to avoid making unacceptable numerals.

Zero – is a difficult concept for children. It can

be introduced as a set with nothing in it, but a situation where there are, for example, two objects such as toy cars and then both of these go away/drive off so that there are 'none' or 'nothing' left is usually more meaningful.

Recognising the difference between sets – for example, of eight and of nine. The eye cannot recognise a set of eight or nine objects laid out in random order, but if each set is laid out in pairs, the set with more is obvious.

WHOLE NUMBERS – PLACE VALUE

What does place value mean?

The method of recording numbers has developed over hundreds of years into a very sophisticated system which has two basic features:

- grouping in tens, and
- using the same digits in different positions for different values.

    There are only ten digits. These are 0, 1, 2, 3, 4, 5, 6, 7, 8 and 9. As well as units, there are an infinite number of values given to these digits – tens, hundreds, thousands etc. These values are all 'powers' of ten (10^1 is 10, 10^2 is 100, 10^3 is 1000, 10^4 is 10 000 and so on). The position of each digit from right to left in a whole number indicates its value, for example:

    345 indicates 5 ones, 4 tens and 3 hundreds
    403 indicates 3 ones, 0 tens and 4 hundreds.

This is called a *base ten place value* system.

What are real life experiences of place value?

Our fingers are grouped into two fives or one ten. This is also true for our toes. It is probably these physical attributes that accounted for man grouping in tens. However, we can make a number system based on groups of any number. Binary is a system based on groups of two where only two digits are required, 0 and 1. These can be given the values – units, twos, fours, eights and so on. One advantage of the binary system is that there are only two options so numbers

expressed this way can be used to control switches being on or off.

What is the key vocabulary and what does each word mean?

The vocabulary includes:

Base – the group, for example, base three would mean grouping in threes. Only 0, 1, and 2 are required as digits and these would have the values ones, threes, nines, twenty-sevens, and so on.

Place value – the position or column in which a digit is written conveys the value of the digit (this is true for decimal fractions as well as whole numbers).

What is the historical background of place value?

Most number systems have developed from the recording of one stroke for one object, for example, the Ancient Egyptians decided to have a symbol to represent ten instead of using ten strokes. They made the first stroke, a linking arc and then the tenth stroke to form a symbol like this

∩

The Egyptians had one of the many systems which were based on grouping in tens. Such systems initially did not have place value and so each new grouping of ten required a new symbol. Some of these symbols were very complex and recording them was laborious, for example, in the Egyptian system three hundred and seventy-six looked like this

Gradually a form of place value was used as a means of shortening the recording, for example, the Romans wrote VI for six, the I on the right of the V means you add it on. However, in time IIII was replaced with IV where I is placed to the left of V and subtracted. The rule became if a symbol for a lesser amount is placed to the left of a symbol for a larger amount it is subtracted from that amount.

The South American Maya people developed a system based on the days in a year and had a base of twenty and a base of eighteen. They are one of the first people to use a symbol like 0 and to have a form of place value. It took many centuries of many different systems before our present system emerged with a base ten grouping and with place value using Hindu-Arabic digits.

What is the value of teaching place value?

It is essential for children to know about and understand the grouping and place value basis of the number system. They must be able to:
- understand that the same digit is used to represent different amounts
- interpret the value of each digit according to its position and find the total amount represented
- express amounts by using the digits.

With these skills the children can communicate numerically.

What are possible key steps in development for the learner?

1 Ten

The children's attention can be focused on the numeral for ten, that is, 10. This is different from the numerals for one to nine as two digits are used, and both the digits already have a meaning when used separately, that is, one unit and zero units. The concept of the '1' having another, different, value from one unit could be explored using materials like Unifix or Base Ten pieces. For example, making a tower or rod of ten ones allows the children to realise that they have created one ten and zero units which can be written as 10. They must also realise that to know whether a 1 is one unit or one ten depends on its position or column.

2 Tens and units

Sets with eleven to nineteen cubes could be looked at as a pile of ones then as one ten and a number of units. The cubes could be placed on a background card (a notation card) which has headings of 'tens' and 'units' at the top of two columns. Digits, each written on a square of card, can be used to label the cubes in each column as shown here:

tens	units
1	3

3 Number names

Number names like fourteen, sixteen, seventeen, eighteen, and nineteen could be discussed so that the relationship between the name and the numeral is found. 'Teen' is explained as a form of ten so that, for example, fourteen is known to be four and ten. *Thirteen* and *fif*teen could be seen as using short versions of three and five along with 'teen'. Eleven and twelve could be explained as remnants from the past before these amounts were considered to be one ten and some units. Ideally this range of numbers should be called ten-one, ten-two, ten-three, to conform to the pattern which is used for the later decades for example, thirty-six, eighty-three.

This study of the teen numbers may be extended to the range of numbers twenty to twenty-nine. Twenty should be accepted as a way of saying two tens, and the children might begin to realise that 2 and probably all the other digits can have the value of tens. In this decade the tens name comes first followed by the unit name, with twenty being used consistently and none of the unit names being shortened. Again, the numbers could be represented by ten towers or rods and units. They could be recorded initially with a heading of tens and units. For twenty to ninety the children will find 'ty' is the short form for tens. They will also find that all these decades conform to the same pattern as the 'twenties'.

4 Hundreds, tens and units

The children should be aware that when they have ten tens they have no single digit to express this in the tens column so they should consider linking the tens together to make 'one hundred'. This creates numerals with three digits and allows 1 and all the other digits to have yet another value. Numbers in the range one hundred to nine hundred and ninety-nine should be explored to learn their names, how to represent them with base ten pieces, and how to write the numeral. Examples like this should be discussed:

375 is 3 hundreds, 7 tens and 5 units,
409 is 4 hundreds, 0 tens and 9 units.

5 Numbers in different forms

The number 46 is most usually thought of as 4 tens and 6 units, but it can also be interpreted as

3 tens and 16 units,
2 tens and 26 units,
1 ten and 36 units, and
46 units.

Children benefit from practice at showing numbers in these different forms, both with base ten pieces and as a written pattern. For example:

numeral	base	ten	pieces
231			

	recording	
2h	3t	1u
1h	13t	1u
	23t	1u
	231u	

6 Thousands

When the children find that ten hundreds are known as one thousand, the base ten pattern of column values should be discussed yet again. Base ten materials could be used once more to represent numbers and to display them in different forms. The thousand is represented as a block which matches ten flats.

7 Hundreds and tens of thousands

At this stage the pupils should realise the repetitive pattern of the column values. They have now met units, tens and hundreds. They have also met 'units of thousands' and are now being introduced to 'tens of thousands' and 'hundreds of thousands'.

They could later be introduced to one million as the name for ten hundred-thousands and find that there is also 'tens, hundreds and thousands of millions'. Pupils could discuss that the number system can represent any value, no matter how large.

What are appropriate resources for teaching place value?

Unifix blocks can be built into towers of ten and cubes kept as separate 'units' or 'ones'. Base Ten pieces, that is, 'cubes' as units, 'longs' as tens, 'flats' as hundreds, 'blocks' as thousands, are ideal to represent the different values because the relationship between the values can clearly be seen. One ten matches ten ones, one hundred matches ten tens and so on. Usually the pieces are set out on a 'notation card' which is an A4 sheet with marked columns for tens and units or hundreds, tens and units.

An abacus can also be used. Here the bead which represents one unit is the same shape and size as the bead which represents one ten. The emphasis is not on relating the different values by size but by position. This is an excellent development to understanding the numeral as 1 looks the same whether it represents units, tens or hundreds etc. It is only its position in the numeral which identifies its value.

What are possible contexts through which place value might be taught?

Place value is not easy to contextualise. However, the context might be of the fantasy type – possibly a special land where the inhabitants are units, tens and hundreds. Money could be considered where 8p is eight pennies, 80p is eight ten-pence coins and 800p is eight pound coins, but most teachers like to keep money as an illustration of the penny as one hundredth of a pound.

Games are ideal here, especially of the type where a player collects units from the throw of one or two dice (where the numbers are added or multiplied). The goal is to exchange the units whenever possible to try to be the first to make, for example, a one hundred piece or a thousand block. There are also games where digits, either written on cards or given on the computer screen, are placed by the player in the order which makes the greatest/smallest number. In the computer version, the digits can appear one at a time for each of two players and the player who can make the number for more/less wins.

How might place value be assessed?

Assess *orally* by asking children to:
- give a value to displays of base ten pieces
- state what number is one more than 29, 99
- state what number is one less than 50, 110, and 300.

Assess *practically* by asking children to:
- show numbers with base ten pieces.

Assess *in written form* by asking children to:
- colour or ring numbers of a stated value
- write in an expanded form, the number of hundreds, of tens and of units.

Assess through *problem solving* asking children to:
- choose a number, then make up clues to help others guess it.

What are common difficulties which children encounter and how might these be overcome?

The principle of place value – This is a very sophisticated concept and it may take many children several years before they can really grasp the principle. Place value of whole numbers benefits from continued revision and consolidation before being applied to decimal fractions. One particular difficulty is interpreting and using zero to represent 0 units, 0 tens etc. A great deal of practice with base ten pieces should help.

The thousand block – For some children the use of the materials causes confusion. Some children thought the thousand block was made up of six hundred cubes because they could see a hundred on each of the six faces. It is essential that the children are involved in exchanging the base ten pieces and in discussing them.

Reading a number – A number such as 123 is read one way, that is one hundred and twenty-three but interpreted in the opposite direction, that is, three units, two tens and one hundred. This can cause confusion unless a great deal of practice is carried out and there is an awareness of this difference.

CONSERVATION

What does conservation mean?

To conserve means, in mathematical terms, to realise that rearrangement of previously matched items does not change them. Young children are greatly influenced by what their eye appears to tell them. For example, a child may set out a blue counter to match each of a given set of red counters.

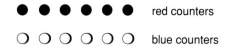

● ● ● ● ● ● red counters

○ ○ ○ ○ ○ ○ blue counters

If the blue counters are then rearranged so that they are spread out and appear as a longer row, children can be deceived by their eye and conclude that there are now more blue counters.

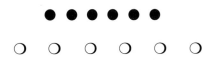

Children who can reason that no counters have been added or taken away, seem to be able to conserve number. However, the test could be carried out using different materials to confirm this result.

  Conservation can apply to measure relationships where rearrangement may mislead children to decide that one of two equal lengths is longer than the other, one of two equal areas is greater than the other, one of two

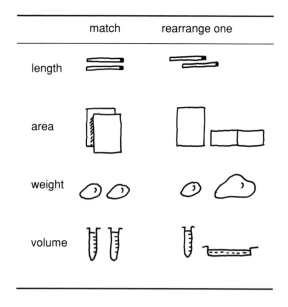

	match	rearrange one
length		
area		
weight		
volume		

What are real life examples of conservation?

In perspective the eye sees a person in the distance as very small but the brain reasons that the person is about as tall as someone nearby.

What is the key vocabulary and what does each word mean?

Comparative terms in many forms are the key vocabulary. For example:

Number

As many as – when members in two sets can be matched in one-to-one correspondence.

16

More – when members in two sets are matched one-to-one and one set has unmatched members.

More/less – when members in two sets are matched one-to-one and one set has unmatched members, this set has 'more', while the other set with all the members matched has 'less'.

Length
As long as – two lengths which are equal.
Longer/shorter – when two lengths are unequal.

Other measures
Area – covers more of the table, covers less of the table, covers the same amount of the table.
Weight – is heavier, is not as heavy or is lighter, would balance.
Volume – holds more than, takes up more space than, holds less than, takes up less space than, holds about the same amount as, takes up about the same amount of space.

What is the historical background of conservation?

Jean Piaget created the term conservation. In his research into how children view their environment he discovered that young children are dominated by what their eyes appear to see rather than by reasoning in number and measure comparison. Children mature to reason through experience and cannot be taught how to conserve.

What is the value of teaching conservation?

You provide activities which should lead children towards achieving conservation. By providing comparison activities involving pairs of sets, children learn to match the members of the sets in one-to-one correspondence and so identify when one set has more than the other and when both sets have 'as many'. Comparison of pairs of objects for the various aspects of measure, give children the opportunity to understand and use relevant vocabulary such as: is longer, as long as, holds more, holds as much as, covers more surface, covers as much surface (table), is heavier, is as heavy as. Manufacturers try to deceive purchasers about weight and volume. They package goods to appear greater than the actual weight and/or volume.

What are possible key steps in development for the learner?

Children seem to grasp conservation of number between the ages of three to five years. Conservation of each of the measures is usually grasped later than conservation of number, with length being the first and volume tending to come after area and weight. Some children aged about ten or eleven may not have matured through their experiences to conserve volume and be inclined to compare two volumes by the height of the container, forgetting that volume is three-dimensional.

What are appropriate resources for conservation?

Children require many practical experiences to grasp conservation. To help them mature and gain conservation of number they should be involved in comparing pairs of sets to find which sets have as many and which have a set with more, using, for example:
- real objects such as chairs, desks, children, books, pencils, cups, saucers
- toys – dolls, cars, zoo and farm animals
- sticky shapes
- drawings.

To help the children gain experience of measure, they should compare pairs of objects:
By length, for example using sticks, pencils, ribbons, string, belts, ties, cardigan back length and sleeve length, coat length.
By area, for example using newspapers, sheets of paper, postcards, envelopes, tablemats, tablecloths.

By *weight*, for example using books, toy cars, pencil cases, parcels, balls.
By *volume*, for example using drinks, bottles, jugs, vases, mugs, cups.

In these experiences the children are not comparing numbers or units of measurement but are making qualitative judgements.

What are possible contexts through which comparison might be experienced?

Play and choosing situations can provide comparison activities. You may also be able to bring these into stories and projects, for example, *The Three Bears*.

How might conservation be assessed?

Any conservation test could have the following basic structure.

1 Making a match
The teacher provides a set of items or an object and asks the child to find (from a given selection) or make, a matching set or object. The child should agree that the two sets have as many or that the two objects are the same length, hold the same amount etc. If the child does not think they are the same you should not proceed with the test but return to practical comparison experiences.

2 Rearrangement
The child is asked to carry out some form of rearrangement, for example, 'put all these red cubes on this plate', 'move the red pencil along here', 'pour the water into this'. If the teacher does the rearranging, the child can be suspicious that some change has been made. If the child does the rearranging he or she should be more able to reason that nothing has been added or taken away.

3 Questioning
The questions should not lead or mislead the

child at this stage. Questions must lay out all the possible options, for example, 'Is the red pencil longer? Is the blue pencil longer? Is the red pencil as long as the blue pencil?'

4 Conclusion
If the child reasons that the sets or objects are still a match, he or she seems to have grasped conservation but it is sensible to repeat the test with different materials or to try another form of the test.

It is usually only necessary to carry out conservation tests with a few children where you believe learning difficulties may be arising from a lack of conservation. However, you may wish to select a sample of children from your class to check this aspect of their maturity.

What are common difficulties which children encounter and how might these be overcome?

Language – Children might not understand the vocabulary, or a previous teacher or adult may have used different comparative terms. Check the child's understanding of the words you use.
Manipulative skills – Children may find it difficult to compare certain objects, for example, flexible lengths like two pieces of ribbon as they may not know how to align two ends and then pull both ribbons through the hand together to show which is the longer. If children are poor at pouring they are likely to spill some as they pour and so amounts cannot be compared for conservation.
Lost among other concepts – Sometimes conservation gets lost among other concepts. For example, some student teachers were exploring an area conservation test. They decided to use two paper squares of equal area for the 'matching' stage. For the 'rearrangement' part of the test, one square was kept whole while the other was halved. One pair of students halved the square into two rectangles, while another pair halved the square

into two triangles. The students then disagreed about the relationship between the area of one rectangle and one triangle. Those who thought the triangle had a larger area must have a lack of understanding about area, could not conserve or their concept of a half was incorrect! However, the test for conservation of the two squares was forgotten!

NEGATIVE NUMBERS

What is a negative number?

Different types of numbers have been devised as they were required. Natural numbers, that is the numbers 1, 2, 3 . . . , were extended to include 0 and were called Whole numbers. Then this set was extended to include numbers less than zero, that is negative numbers. All these numbers, . . . $^-3, ^-2, ^-1, 0, 1, 2, 3$ . . . , are called *Integers*.

Negative numbers are defined also as the answers when whole numbers are subtracted from 0, for example, $0 - 9 = {}^-9$. Negative numbers occur in any subtraction calculation where the number subtracted is greater than the number it is subtracted from, such as $2 - 3 = {}^-1$. Note how the negative sign is written much higher than the subtraction sign to show clearly that negative one ($^-1$) and minus one (-1) are different. When there is no sign the number is interpreted as positive.

Negative numbers can also be called *directed* numbers and in this context they are used to indicate direction. On a horizontal scale movements to the right of a fixed starting point, which is labelled 0, are represented by positive numbers and movements to the left by negative numbers. On a vertical scale movements downwards from the fixed point are regarded as negative.

What are real life examples of negative numbers?

Temperature can be expressed below zero in cold weather. Such temperatures should be referred to as 'negative' not 'minus' as you sometimes hear in weather reports. Negative amounts in a bank account are referred to as being overdrawn or 'in the red'. Graphs can be drawn with negative numbered axes and have negative coordinates.

What is the key vocabulary and what does each word mean?

Positive – greater than 0 and in the direction of increase.
Negative – less than 0 and in the direction of decrease.
Number – there are different uses of number; cardinal number states 'how many' members are in a set whereas ordinal number states the position of one member in a set.
Number line – integers (positive and negative numbers) expressed at directed regular intervals on a line with zero as a fixed point.
Minus – subtract.

What is the historical background of negative numbers?

Negative numbers were used by many early mathematicians, for example there is evidence of their use by the Chinese in the second century BC. The Hindus represented negative quantities with a dot. However, a copy made about the ninth century of an older document was discovered in a town on the northern Indian frontier and the sign used there for a negative quantity was a +. In the eighteenth century the Italian Cardan made use of directed numbers which was considered a modern concept of the use of negative numbers. Napier in the early seventeenth century regarded negative numbers both as 'less than nothing' and as 'representing a point moving in a direction opposite to positive'.

What is the value of teaching negative numbers?

Using negative or directed numbers allows more situations to be expressed numerically.

What are possible key steps in development for the learner?

1 The concept
The use of a lift as a context might give some meaning to negative numbers. The ground floor can be regarded as floor 0 with the floors above as positive and the floors below like the 'lower ground' and 'basement' as negative. Another context is winter where the use of a thermometer allows learners to observe decreases which fall below 0°C and require to be represented by negative numbers.

2 The notation
The numeral should be preceded by a negative sign written near the top of it, for example, $^-18$. Learners should realise that this does not mean subtract but explains that either the number is less than 0 or its direction.

3 Computation
Children in primary school are not usually expected to handle negative numbers abstractly in operations. The emphasis should be on rises and falls, increases and decreases, moves up and down or forward and back, which involve negative numbers and these are best found using a number line. This is likely to involve, for example:

- additions arising from word problems like these:

Sharon enters the hotel lift at the beach level which is labelled $^-3$. She goes up two floors. What level does she get out at?

$$^-3 + 2 = {}^-1$$

Aziz had borrowed £3 from his Mum. He wrote $^-3$ in his notebook. He borrowed another £2. What calculation would he write in his notebook to show the total amount he now owed her.

$$^-3 + {}^-2 = {}^-5$$

- subtractions arising from word problems like this:

The temperature was $^-3$°C last night. During the night it fell two degrees. What temperature did it fall to?

$$^-3 - 2 = {}^-5$$

More difficult situations, for example those which might lead to subtracting a negative number which results in adding the number, is best left till the children are much older, so do not attempt calculations such as $^-3 - {}^-2 = {}^-1$.

What are appropriate resources for teaching negative numbers?

Vertical and horizontal number lines, a thermometer, weather reports which include temperature.

What are possible contexts through which negative numbers might be taught?

Weather, holidays, coordinates, graphs.

How might negative numbers be assessed?

Assess *orally* by asking children to:
- explain in their own words what a negative number is

- read a negative number.

Assess *practically* by asking children to:

- read temperatures which are below zero.

Assess through *written examples* such as:

- William enters the lift at floor 3. He goes down four floors. What number would you use for the floor he has arrived at?

Assess through *problem solving*:

- For a class race, give runners starting positions from ⁻3 to 3 based on trial performances.

What are common difficulties which children encounter and how might these be overcome?

The concept – Children find it difficult to understand zero because it represents, for them, something which does not exist. Numbers which represent quantities less than zero also represent the non-existent for many young children and so are likely to pose problems for many of them. It is sensible to consider very carefully if certain children are mature enough to be introduced to negative numbers. When the introduction is made, it should be through practical situations. This is likely to be the use of negative numbers indicating direction, either 'down' or 'back'. The understanding of 'less than zero' as a negative value can come later.

Calculations – It is sensible to do this work only based on practical examples and only with some children.

COMMON FRACTIONS

What is a common fraction?

A fraction is a part of a whole. When the whole or unit is divided into equal parts you have a 'mathematical fraction'. The description 'common' means simple or ordinary.

Depending on the number of equal parts, each is given a name, for example:

- where there are two equal parts, each is one half
- where there are three equal parts, each is one third
- where there are four equal parts, each is one quarter
- where there are ten equal parts, each is one tenth
- where there are fourteen equal parts, each is one fourteenth
- where there are sixty-four equal parts, each is one sixty-fourth.

The most common method of naming the part is to use the number name with 'th' added to the end of it.

The notation used is to show the number of equal parts as 'a number below a line' or denominator, for example, one half is recorded as $\frac{1}{2}$ and one quarter as $\frac{1}{4}$. The number above the line, the numerator, signifies the number of parts being referred to, for example, $\frac{3}{4}$ indicates three of the four equal parts. All whole numbers can be expressed in this notation, for example:

$$1 = \tfrac{1}{1}, \tfrac{2}{2}, \tfrac{3}{3}, \tfrac{4}{4}, \tfrac{5}{5} \qquad \text{and so on}$$
$$2 = \tfrac{2}{1}, \tfrac{4}{2}, \tfrac{6}{3}, \tfrac{8}{4}, \tfrac{10}{5}, \qquad \text{and so on}$$
$$18 = \tfrac{18}{1}, \tfrac{36}{2}, \tfrac{54}{3} \qquad \text{and so on} \ldots$$

Any number which can be expressed as a numerator over a denominator is called a *rational number*. All common fractions are rational numbers. All integers are rational numbers.

What are real life examples of common fractions?

Measurements are often expressed as fractional parts of main units, for example, at the supermarket there are goods sold by the half kilogram (sugar, salt, butter, margarine), and the half litre (milk, lemonade, olive oil, cooking oil).

Material might be bought to the nearest half or the nearest quarter metre, for example $1\frac{1}{4}$ metres for a skirt, or $10\frac{1}{2}$ metres for curtains. Time is expressed as quarter past, half past and quarter to the hour. However, fractional parts of metric measures are mainly expressed as decimal fractions, apart from halves and quarters so common fractions are not widely used.

What is the key vocabulary and what does each word mean?

Mathematical fraction – an equal part of a whole or an amount.
Common, vulgar or proper fractions – when the whole is divided into any number of equal parts.
Denominator – the number of equal parts expressed below a horizontal line in notation.
Numerator – the number which indicates the number of equal parts being referred to and is expressed above the horizontal line in common fraction notation.
Equivalent fractions – the same value expressed in different forms, for example $\frac{1}{2}$ and $\frac{2}{4}$, $\frac{3}{4}$ and 0.75.
Mixed numbers – a whole number alongside a common fraction, for example $2\frac{1}{2}$ which expresses an amount as: 2 wholes and 1 half of a whole, and/or greater than 2 but less than 3 with the accuracy of 'to the nearest half unit'.
Improper fraction – a mixed number can be expressed as a numerator and a denominator, for example $2\frac{1}{2}$ is the same as $\frac{5}{2}$; in improper fractions the numerator is larger than the denominator.

What is the historical background of common fractions?

Measurements sometimes needed to be expressed with greater accuracy than whole numbers would allow, for example the length of a stick may be expressed as:

$$\text{between 1 and 2 m, or}$$
$$\text{about } 1\tfrac{1}{2} \text{ m, or}$$
$$\text{about } 1\tfrac{3}{8} \text{ m or}$$
$$\text{about } 1\tfrac{5}{16} \text{ m and so on} \dots$$

Each of these measurements is more accurate than the preceding one.

The Egyptians used simple fractions and had notation for halves, thirds, quarters and sixths. Greek writers often expressed fractional values in words, for example Archimedes expressed the length of a circle as three diameters and part of one, the size of which lies between one-seventh and ten-seventy-firsts. The Hindus wrote fractions with a denominator beneath a numerator. An Arabic author of the twelfth century, al-Hassar, seemed to be the first to comment on the line used in notation by instructing that denominators be written below a horizontal line.

It was, however, about the close of the fifteenth century that the present notation $\frac{1}{2}$ appeared.

What is the value of teaching common fractions?

Some common fractions are still used, so children should be given the opportunity to develop a concept of a fraction, and be able to express their understanding in words and notation. Fractional work is, of course, linked to division and children should realise that, for example one half of an amount is the same as dividing it equally between two, and that one quarter of an amount is the same as one of four equal shares.

What are possible key steps in development for the learner?

1 Recognising and creating equal parts of a whole
Fortunately the words 'half' and 'quarter' may be used in the home as items are shared, so the children are likely to have heard them in a meaningful context. They may even use them as part of their own vocabulary. Young children can create equal parts by folding lengths of string or ribbon, and paper shapes such as the rectangle and circle. It is during such activities that the fraction names should begin to be associated not only with a number of parts but with a number of equal parts.

2 Language
The children should talk about practical activities in terms of 'I am folding this shape to make four equal parts. Each of the parts is called one quarter.' The generalisation of how to 'name' denominators is best delayed until the learner can differentiate between 'sixth' as one of six equal parts and 'sixth' as the position of one member of a group (the ordinal use of number).

3 Practical experiences
Halves and quarters are initially the result of a practical activity like folding. It is good to continue to give practical work to older children, possibly as a problem to solve, for example 'give Scott half of this piece of Plasticine', 'halve this bag of pasta shells', 'put half of this bottle of water in the jug'.

Fractions can be recognised and created on drawings of shapes. When children first meet the half of an amount like 6, they may tackle this by setting out an array and then halving this as if it was a shape, to find an answer of 3.

```
      *   *   *
    ---------------
      *   *   *
```

Later the equivalence of 'one half of' with 'divided by 2' and 'one quarter of' with 'divided by 4' can be investigated.

4 Equivalence
Children could learn by using a number strip, Cuisenaire rods or sectors of a circle, for example, that two quarters are the same as one half that four quarters are the same as one whole, that five quarters are the same as one whole and one quarter.

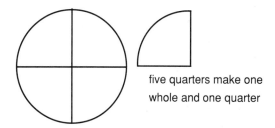

five quarters make one
whole and one quarter

5 Computation
Children could learn:
- to add and subtract simple fractions like halves and quarters
- to find one half of, one quarter of, and three-quarters of amounts
- to find how many halves or quarters there are in a mixed number.

6 A common fraction as a rational number
The generalised concept that the whole can be made into any number of equal parts, and that this can be recorded as a denominator, should be realised in time by most children. They should come to terms with the fact that the larger the number of parts (denominator), the smaller the parts.

Eventually a fractional part should be seen as a number which has a position on a number line and on measurement scales, at first between 0 and 1 and then related to other whole numbers and to integers. At this stage in the development common fractions should be considered as abstract numbers (rational numbers) which do not require to be related to real objects or to measurement, for example, that $2\frac{1}{2}$ is a number greater than 2 but less than 3.

What are appropriate resources for teaching common fractions?

Cuisenaire rods, shapes divided into equal parts, lengths of flexible material like string and ribbon, paper shapes for folding, squared paper to colour equal parts of shapes, a number line, scales for length, weight and capacity.

What are possible contexts through which common fractions might be taught?

Shopping, the supermarket, sharing in the family, twins, triplets, quadruplets . . .

How might common fractions be assessed?

Assess *orally* by asking mental calculations like:
- Which is greater, one half or one third?
- Which is smaller, one thirteenth or one fifteenth?
- What is one half and one quarter? What is one and a half and three-quarters?
- What is one half of twenty-four?
- How many quarters are there in two and a half?
- Steve's Dad has bought two bottles of lemonade. One bottle holds one and a half litres, the other holds two litres. How much lemonade has he bought altogether?
- The jug holds half a litre. How will Sharon use the jug for:
  (*a*) one quarter of a litre?
  (*b*) three-quarters of a litre?

Assess *practically* by asking children to:
- make halves or quarters of items or amounts.

Assess through *written examples* which involve:
- colouring half of a shape
- drawing a shape with half of the area of a given shape

- a calculation such as: 'The petrol gauge in Mr Shearer's car shows that it is half full. When the tank is full it holds 85 litres. How much petrol does it hold just now?'

Assess through *problem solving* by asking children to:
- choose a set of rods from this box (Cuisenaire or Colour Factor) which you can label as one whole, one half and one quarter; choose a different set which could have the same labels
- use one sheet of A4 paper to make a visual aid to show the fractions – one half, one quarter, one eighth and so on . . .
- make up a bag with one quarter of these pasta shells in it.

What are common difficulties which children encounter and how might these be overcome?

Understanding the notation – Some children find it difficult to grasp that the larger the number of equal parts, the smaller each part is, and that the larger the denominator the smaller the fraction. They should benefit from making fraction parts with a wide range of materials.

Same name fractions – Common fractions should be compared by changing them to the same name, that is, the same denominator. Any difficulty in understanding here might be helped by the children being given two sets of paper shapes marked one whole, one half, one quarter and one eighth with the task of making one set of the pieces into eighths. The eighths are then used to fit on top of the uncut shapes and results like 'eight eighths are the same as one whole', 'four eighths are the same as one half' and so on. The children should be aware that common fractions are changed to the same name before being added or subtracted and should use materials to show this.

DECIMAL FRACTIONS

What does decimal fraction mean?

A common fraction can be thought of as the whole divided into any number of equal parts, so a decimal fraction is the whole divided into ten, one hundred, one thousand etc. equal parts. These fractions, in which the number of equal parts is a power of ten, can be recorded by extending the number system. For example, 3·4 means three units and four tenths of a unit. In whole numbers, for example, 34, the last digit on the right is the units digit. If the place value columns are extended to the right then a marker is required to indicate the units digit. A dot is used as this marker.

Decimal fractions are another method of expressing some common fractions, so decimal fractions are also *rational numbers*.

What are real life examples of decimal fractions?

Tenths are to be found in systems of measure. For example, metric measures for length and volume use fractions like one tenth and one hundredth and one thousandth to relate the units. A centimetre is one hundredth of a metre and a millilitre is one thousandth of a litre.

What is the key vocabulary and what does each word mean?

Decimal – powers of ten and/or a base of ten.
Tenth – one of ten equal parts.
Hundredth – one of a hundred equal parts.
Thousandth – one of a thousand equal parts.
Decimal point – a dot used to mark the units as being the digit to the left.

What is the historical background of decimal fractions?

Michael Stifel, a convert of Luther, is credited with extending the framework of our place value system 'to the right'. At first, a gap was left after the units digit but later a dot was used.

What is the value of teaching decimal fractions?

Money and all the measures except time have a base of ten and so can be recorded as decimal fractions, for example £3·68, 4·835 litres, 7·82 m, 1·6 cm^3.

What are possible key steps in development for the learner?

1 Revision of place value for whole numbers
The focus should be on the relationship of the columns, that is, each column to the left is ten times greater, and each column to the right is ten times smaller. This can be illustrated by looking at the unit cube, the ten 'long', the hundred 'flat' and the thousand 'block' of Base Ten structured pieces.

2 Introduction of tenths
Usually children are introduced to materials and language to establish the 'new' concept. In this instance the concept of one tenth should have been met as a common fraction (one tenth is one of ten equal parts), so the emphasis can be on another form of recording one tenth. One approach is to extend the whole number place value notation by considering that one ten is one tenth of one hundred, one unit is one tenth of one ten, so it seems simple to create a new column on the right for tenths of one unit.

3 Representations of tenths
Base ten pieces can now be used with the 'block' which usually represents one thousand, being labelled to represent one unit. The children would find that ten 'flats' make a block so each is one tenth. Pieces could be displayed to represent numbers like the ones shown on this

page: one unit and 3 tenths or 1·3, like this

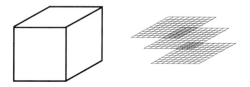

If it is felt that this change in use of the Base Ten pieces might be confusing, use something like a card rectangle, 20 cm by 5 cm, for the unit and another rectangle, 5 cm by 2 cm for the tenth. Pieces would then be for one unit and 3 tenths or 1·3.

4 Measuring to the nearest tenth
A metre stick could also be used to represent one unit. The children should find that ten orange Cuisenaire rods match one metre, so each is one tenth. Metre sticks and orange rods can be used to measure lengths to the nearest tenth of a metre. The following diagram shows the tiddlywink is 2 metres and 3 tenths from the start (2·3 m).

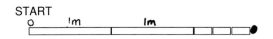

The volumes of containers can be found using a litre and a tenth of a litre (100 ml), for example the volume of the vase is 2 litres and 1 tenth (2·1).

The children are likely to be interested in seeing the odometer in a car, especially when it is recording tenths of a mile (or kilometre) as the car moves along the road. Some digital stop watches show tenths of a second, but this could be confusing to many as hours, minutes and seconds are not expressed in decimal notation.

6 Extension to hundredths
Use:
- the Base Ten pieces with the 'long' as the hundredth to represent numbers with units, tenths and hundredths
- 1 penny as the hundredth of £1 to show amounts like £2·72
- metres and centimetres for measurement of lengths.

The children should have practice both at representing, recognising and recording in decimal notation amounts and measurements. It is important that the pupils should realise that a number can be expressed in different ways, for example:

3·67 is 3 units, 6 tenths, 7 hundredths
    and   36 tenths, 7 hundredths
    and   367 hundredths
    and 3 units   67 hundredths
    and so on . . .

7 Extension to thousandths and other decimal places
The thousandth can be illustrated through measures like one millilitre as 0·001 litre, one millimetre as 0·001 metre. A few measurements should be made to the nearest millilitre (millimetre) and recorded as litres (metres) with three decimal places.

The calculator shows up to seven decimal places and can be used to give children experience in rounding to the nearest whole number, to a specific number of decimal places and to a specific number of significant figures. For example:

3·7654263 is 4 to the nearest whole number
          is 3·8 to the first decimal place
          is 3·8 to two significant figures.

7 Rules for rounding a 5

A 5 can be rounded to the nearest even number, for example, 3·5 is taken as 4 to the nearest whole number, 6·45 is considered as 6·4 to the first decimal place. However, a simpler rule is to round a five to the higher number, for example, 6·45 would become 6·5 to the nearest tenth.

8 Calculations with decimals

These are considered in this book under the headings of the different operations. They are not just whole number calculations with 'dots'. Each operation involving decimals requires to be:

- introduced through materials
- has new language to be learned
- requires graded examples so that one new concept or skill is introduced at a time.

What are appropriate resources for teaching decimal fractions?

Base Ten structured material can be used with each piece representing a different value from those adopted for whole numbers, for example, a 'block' could represent a unit, the 'flat' a tenth, the 'long' a hundredth, and the 'cube' a thousandth. A metre could be the unit with a ten centimetre rod for one tenth and a centimetre cube for a hundredth.

Money can be used with a pound coin as the unit, a ten pence coin as the tenth and a penny as the hundredth. However, it is not sensible to use money to represent pounds and tenths as £1·6 is *not* an acceptable recording for one pound and six ten pence coins.

What are possible contexts through which decimal fractions might be taught?

'Ourselves' where length measurements for heights, waists, arms etc. are taken and recorded with decimal fractions of a metre.

'Shopping' where prices, for example, of clothes, are labelled, then customers buy items. This context could make use of catalogues where buyers use forms for home orders.

How might decimal fractions be assessed?

Assess *orally* where children:

- explain in their own words what a decimal fraction is (a tape-recorder could be used with a group listening to all the individual contributions and deciding who gives the best explanation)
- interpret decimal notation and explain the value of the different digits.

Assess *practically* by asking children to:

- use number pieces to represent the number 2·106
- measure the height of a classmate and express the answer in metres to the nearest hundredth
- measure the volume of a vase to the nearest 10 millilitres and express the answer in litres.

Assess *in written form* where children:

- record lengths, volumes and prices using decimal notation
- 'round' decimal notation, either measure results or answers to calculations.

Assess through *problem solving* by asking children to:

- design and make a decimal display which can be used to show classmates what happens to the digits in a decimal fraction when it is multiplied by ten or a hundred.

What are common difficulties which children encounter and how might these be overcome?

Concept of a decimal fraction – Some children think the point is the decimal fraction, lacking understanding of place value. They require more experience of explaining the value of each of the digits for numbers when there is decimal notation. Others forget to use the point when recording a decimal fraction and don't realise that they have written a whole number and not what they intended.

Value of numbers where decimal notation is used – The magnitude of numbers can be confusing, for example, when asked 'Put these numbers in order, smallest first: 1, 1·01, 1·11, 0·01, 0·1', many children would be uncertain about how to do this. It may be necessary to help them find a method of comparison. For example, some children could think of each as a number of hundredths, that is, 100, 101, 111, 1

and 10, while others might be able to compare a few at a time, that is, 1·01 and 1·11 are more than 1; 1·11 is more than 1·01; 0·1 and 0·01 are both less than 1; 0·01 is less than 0·1.

Zeros – may cause confusion and again the children require practice in giving values to each of the digits. Where there are decimal fractions with a different number of places to be added, for example, 0·5, 3, 0·23 and 6·109, children need to be careful to note where the unit digit is in each. In addition and subtraction calculations, the unit digits and the points should be aligned.

PERCENTAGES

What does percentage mean?

Common fractions can be used to express any number of equal parts of the whole. Decimal fractions only express numbers where the equal parts are powers of ten like tenths, hundreds and thousandths. Percentage fractions are even more limited as they are used only to express the whole divided into one hundred equal parts. One percent means one hundredth part of a whole or unit, and the notation used is 1%. Percentage fractions are used for comparison. Many fractions are difficult to compare, for example, to compare $\frac{4}{5}$ and $\frac{3}{4}$ the fractions are changed to equivalent values with the same name. One suitable 'same name' for fractions is hundredths, for example, $\frac{4}{5}$ is 80 hundredths and $\frac{3}{4}$ is 75 hundredths. Percentages, as another form of common and decimal fractions, are rational numbers.

What are real life examples of percentages?

Many increases and reductions in amounts are expressed as percentages, for example, employees could be awarded a 10% rise in wages, a sale offers reductions of 15%, and

manufacturers of a product like shampoo will sometimes offer us 5% more for the same price.

What is the key vocabulary and what does each word mean?

Percent – per hundred.
Percentage increase/decrease – the amount more/less expressed as hundredths, for example: What is the percentage increase in price if a book had previously cost £3 and was now labelled £3.24?

Increase is 24p
Previous price was 300p
Fractional increase is $\frac{24}{300}$ or $\frac{8}{100}$ or 0.08
Percentage increase is 8%

What is the historical background of percentages?

An Italian writer of about 1425 used a symbol which is the basis of the present notation. Instead of writing Pc for per cento he put $P\frac{c}{o}$. This became per $\frac{o}{o}$ about 1650 and gradually the 'per' was dropped and the line between the zeros slanted to give the symbol used today.

Percentages have led to two common sayings. 'Fifty-fifty' refers to 50% and means that something should be halved between two people, for example, the cost of a fence between neighbours. 'There is no percentage in it' refers to profit which is usually expressed in percentage terms and means there is no advantage for the speaker in doing the task.

What is the value of teaching percentages?

Percentages are regularly used as a form of communication about fractions. This may be because so many measures and money are decimalised, for example:

1% of £1 is 1p
1% of 1 metre is 1 centimetre
1% of 1 litre is a centilitre or 10 millilitres.

If wages are to be raised by 6%, this can be thought of as 6p in every £, or £6 in every £100 which can be understood in concrete rather than abstract terms.

What are possible key steps in development for the learner?

1 Understanding the concept
The child should be able to recognise a whole which has been divided into one hundred equal parts, for example, a square drawn on centimetre squared paper with an edge length of 10 cm is made up of one hundred centimetre squares, a metre is made up of 100 cm, and £1 has the same value as 100p. The child should be able to produce his/her own hundredth parts of a whole, for example, by drawing a rectangle 20 cm by 5 cm on centimetre squared paper, each centimetre square representing one hundredth part of the rectangle.

2 Notation
'One hundredth' is the value for both of these written recordings 0.01 and $\frac{1}{100}$. The percent

symbol % used with 1 is simply another format for recording the same fraction. For example, all these expressions represent the same fraction:

twenty-three hundredths $\qquad$ $\frac{23}{100}$
0.23 $\qquad$ 23%

The percentage notation is considered the easiest to write because it can be expressed most of the time as a whole number. However, this possibly makes the meaning more difficult if the concept of percent is not understood.

3 Calculations – finding a percentage of an amount
One percent is one hundredth of any amount or the amount divided by 100, for example:

1% of 180 is 180 ÷ 100 or 1.8

To find a greater percentage, it is possible to consider finding 1% mentally by a place value move of the digits and then to multiply the result by the required percentage, for example:

15% of 180 is 15 × 1.8 = 27

Of course, the calculator may be used for the multiplication. You can also use the % key. Calculators vary but the usual method is like this:

| 1 | 8 | 0 | × | 1 | 5 | % | gives | 27 |

Note that the = sign should not be used.

4 Calculations – expressing a fraction as a percentage
We can arrange the 'whole' amount to be one hundred so that hundredth fractions can be easily understood. For example, if we want to know 40% of £1, 1% is one penny, so 40% is 40 pence.

However, in many real situations the fraction is not expressed 'out of one hundred' but is in a different form. For example an interviewer talks with 48 people and finds that 18 use the product being investigated. The fraction of people using the product can be expressed as:

- a common fraction with a numerator and denominator, that is $\frac{18}{48}$, or $\frac{3}{8}$ in its simplest terms
- a decimal fraction by dividing the numerator by the denominator to produce 0.375
- a percentage by multiplying the common fraction or the decimal fraction by 100, that is 37.5% or $37\frac{1}{2}$%.

  The calculator % key may of course be used. The above example would be entered as:

1 8 ÷ 4 8 % giving 37.5

What are appropriate resources for teaching percentages?

Squared paper, the Base Ten flat (which represents one hundred units) and unit pieces, a metre stick and centimetre cubes, newspaper and magazine cuttings with references to percentages.

What are possible contexts through which percentages might be taught?

Shopping, the supermarket, holidays, the school tuckshop, selling our home-made jewellery (cakes/sweets etc.), newspapers.

How might percentages be assessed?

Assess *orally* by asking a group:
- about their understanding, for example:
  What does percent mean?
  What does 15 percent mean?
- some simple calculations:
  What is 1% of 500?
  What is 10% of 60?
  What is 0.16 as a percentage?
  What is a mark of 4 out of 10 as a percentage?

Assess *practically* by asking children to:
- carry out a survey to find the time individuals spend watching TV in a week, then display the results as a percentage of the number surveyed.

- cut a length of ribbon, for example 20 cm, and then another piece which is 25% longer
- pour out a volume of water, for example 600 ml, and then another volume which is 20% more than that
- weigh an amount, for example 800 g of pasta shells and then another amount which is 15% less.

Assess *in written form* by presenting children with questions like:
- Sharon can type 70 words a minute. After practising she can type 77 words a minute. What is her percentage increase?
- A mountain bike costs £450. VAT at $17\frac{1}{2}$% is added. What is the total cost?

Assess through *problem solving* by asking children to:
- draw up a list of 'percentage challenges' for your group to meet, for example:
  to take 10% more exercise per week
  to watch 5% less TV per week.

What are common difficulties which children encounter and how might these be overcome?

Calculations without understanding – Carry out examples with practical materials. For example, to find 5% of 300:
- use 3 flats to represent 300
- 1% of 300 is one hundredth of 300 (1 unit cube for each hundred flat), that is 3 (the short way of doing this is to divide by 100)
- to find 5% of 300, the learner puts out five lots of 3 to get 15 (multiplication by 5).

Percentages greater than 100 – Some children find it difficult to accept percentages like 125%. An explanation could be based on a shop selling an item which has been bought for £30. A 50% profit would mean it is sold for £30 plus £15, that is £45. A 100% profit would mean it is sold for £30 plus £30, that is £60. A 150% profit would mean it is sold for £30 plus £45, that is £75.

THREE

Using numbers

ADDITION

What does addition mean?

A number is the cardinality or 'how-many-ness' of a set. When two separate sets of objects are put together to form a third set this is an illustration of adding two numbers.

In set language, addition is expressed as the 'union' of two 'disjoint' sets. Addition is a procedure carried out with two numbers and this is an example of a 'binary operation'.

What are real life experiences of addition?

People can, for example:
- add coins to pay for an item
- add prices to find the total spent
- add time to find a finishing time
- add measurements to find the total amount of cloth required
- add the number of children in each class to find the total school role.

What is the key vocabulary and what does each word mean?

Add, and, plus – all these words are used to mean to *combine* numbers or quantities.
Total, sum – refer to the unique answer when numbers or quantities are added.
Binary operation – a procedure which involves two numbers.

What is the historical background of addition?

Addition was initially carried out as a count and a counting frame or abacus was used. The Egyptians used the symbol of a pair of legs walking from right to left, the direction of Egyptian writing, for addition (although a later manuscript uses the same symbol to square a number). Some universality emerged in the fifteenth and sixteenth centuries when mathematicians began to use 'p' to represent plus. The modern + came into use in Germany towards the end of the fifteenth century. Around 825 AD, a Persian mathematician called al-Khowarizmi wrote about rules for setting down basic calculations. Today the step-by-step method which is used to carry out a calculation like an addition is called an algorithm after him.

What is the value of teaching addition?

Addition is regarded as a basic calculation skill which has a value for recording and communicating. Addition can be carried out by counting, but children are encouraged to memorise basic facts. Addition involving the same number leads to multiplication. Subtraction can be carried out by adding to the smaller number.

What are possible key developments for the learner?

1 Counting on

The first introduction to addition is usually through counting on to find one more, for example, three and one more is four. This is then extended to counting on two more. Here some children find it difficult to count on and use the process that they use for most addition facts. For example, to add 3 and 2, the child counts out three, then counts out two and finally counts from one to five to arrive at the sum. Children require help and encouragement to start at the first number and count on. Towers of linked Unifix blocks or a number strip or line can encourage this, for example:

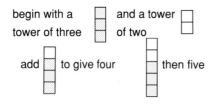

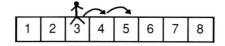

stand on 3 on a number strip

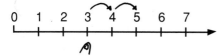

put a finger at 3 on a number line

In each instance the child might say 'To add three and two I start with the "three" and I say "four", "five". Three and two are five.'

2 Memorising facts

Children can guess the sum, then check by matching Unifix towers or Cuisenaire rods. They could then be involved in practice games

to memorise the facts. Each of the following steps should be carried out with a few facts at a time:

- counting each set separately, then the joined sets, counting from one to find the total
- counting on from the first set
- guessing the sum and checking with materials
- getting to know the fact.

For example, for the numbers 1 to 5, the facts could be 'parcelled' as:

- adding one $1+1, 2+1, 3+1, 4+1$
- adding two $1+2, 2+2, 2+3$
- adding three $1+3, 2+3$
- adding four $1+4$

These facts could then be revised as 'stories' like this:

- story of 3 $1+2, 2+1$
- story of 4 $1+3, 2+2, 3+1$
- story of 5 $1+4, 2+3, 3+2, 4+1$

3 Facts involving zero

Adding zero, that is a set with nothing in it, is difficult for young children. Mathematicians call this the identity element for addition, as the other number in the operation remains unchanged. Children need practice with examples where zero is involved.

4 The commutative property of addition

Children can discover that it does not matter in which order sets are added as the total remains the same. If they find $1+4$ difficult, they soon realise that $4+1$ is much easier. Mathematicians call this the commutative property of addition. If children accept that order is not important it greatly reduces the number of facts they need to memorise.

5 Facts with a sum equal to or less than 10

Some teachers teach the facts to 5 and then to 10. It is certainly beneficial to children to learn only a few at a time. Encourage children to apply their knowledge of addition facts in the context of classroom experiences and theme work.

6 Facts with a sum equal to or less than 20

Although these facts can be learned in the same way as those up to 10, many teachers prefer to use the facts already learned to advantage. This means realising numbers 10 to 19 are one ten and some units. The teacher, for example, to add $15 + 4$, expresses the first number as $(10 + 5) + 4$ and makes use of the associative property by carrying out the second addition first, that is, $10 + (5 + 4)$. As far as the children are concerned they are seeing the numbers in tens and units columns. By using Base Ten pieces on a notation card, the children soon realise that all they are doing is revising the story of nine and extending it to include the fact 'zero tens plus one ten'. The story of 19 would be recorded like this:

1	2	3	4	5	6	7	8	9
$+18$	$+17$	$+16$	$+15$	$+14$	$+13$	$+12$	$+11$	$+10$

For the story of 18, there is one new fact to be learned, $9 + 9$. For 17, two new facts $9 + 8$ and $8 + 9$. For 16, three new facts, $9 + 7$, $8 + 8$, $7 + 9$, and so on. These facts could introduce children to the process of exchanging ten units for one ten.

7 Adding tens and units

Here the children add units and then add tens, initially using tens and units pieces on a notation card. The process of exchanging ten units for one ten and recording this with a crutch figure is the main teaching point. However, it is important that children learn to explain the algorithm. They could adopt language like this:

$$
\begin{array}{r}
45 \\
+38 \\
\hline
83 \\
\hline
{\scriptstyle 1}
\end{array}
$$

I have to add forty-five and thirty-eight.
Forty-five is four tens and five units.
Thirty-eight is three tens and eight units.

Add the units. Eight and five are thirteen.
Thirteen units are one ten and three units.
Add the tens. Three and four are seven and one more is eight tens.
The sum is eight tens and three units or eighty-three.

8 Adding mentally

Many children could use their memorised facts to advantage for additions like $34 + 5$, where they remember that $4 + 5 = 9$ so $34 + 5 = 39$. Adding ten is easy so adding nine might be thought of as 'adding ten then subtracting one', so $45 + 9$ would be $55 - 1$ or 54.

9 Adding more than two numbers

Children should realise that they are only able to add two numbers at a time and then add the sum to another number, so $6 + 9 + 7$ is calculated as $7 + 9 = 16$ then $16 + 6 = 22$.

only able to add 2 numbers at a time

10 Adding numbers regardless of magnitude

When first adding hundreds it is best to use the Base Ten materials again as the exchange of ten tens for one hundred is a different language from ten units being the same as one ten. This is also true for thousands and the exchange of ten hundreds for one thousand. After these experiences, children should realise the exchange of 'ten for one' is a consistent relationship in the place value system.

11 Word problems

Children should be able to recognise addition calculations when they are presented in the context of word problems. The word 'altogether' should be recognised as one indicator of addition.

12 Calculators

Simply adding on a calculator is very dull. However, using the calculator can be fun for addition patterns. Children could learn to round numbers and find an approximate answer to additions then use the calculator to find the exact answer.

Calculations can also be checked by repeating the addition using the numbers in a different order. The calculator is best used when the child is solving a problem where, for example, finding the totals of different numbers might allow an easy check before deciding which path to follow towards a solution.

What are appropriate resources for teaching addition?

Discrete materials for the early stages can be related to a range of contexts, for example, animal shapes, Christmas presents, pebbles from the beach. Structured materials like Unifix which can be both discrete and continuous are widely used. Cuisenaire is not so popular but is useful for stories of the numbers to ten. Base Ten pieces are frequently used and are excellent as an aid to children's understanding of steps of the process. The number strip is useful at the early stages and a number line can be used for numbers up to 100. Addition of money could involve coins and bank notes. Addition of time could involve clocks and timetables.

What are possible contexts through which addition might be taught?

Games at the fun fair could provide opportunities for addition as could many other contexts like the circus, the holiday, ourselves.

How might addition be assessed?

Assess *orally* by asking children to:
- carry out mental additions
- give a different two numbers which add to the same total as $3 + 6$
- give pairs of numbers which add to 24
- explain what 'to add' means
- talk their way through a calculation like

$$345$$
$$+558$$
_____

_____

Assess *practically* by asking children to:
- use Base Ten pieces to carry out a calculation like

$$47$$
$$+78$$
_____

_____

- find the sum of a selection of coins
- use a time line to add 1 hour 35 minutes to 11:10 am
- use a calculator and two dice (one shows tens and the other units) to add numbers until a sum over 400 is obtained.

Assess *in written form* by asking children to:
- add a set of about eight graded examples
- make up four addition examples for a classmate to do
- find the answers to four addition word problems
- identify which word problems from a given set of ten lead to an addition calculation
- make up two addition word problems for a classmate to do.

Assess through *problem solving* by asking the children to:
- find five different pairs of numbers where each pair add to make 82
- think of a number then find two different sets each of three numbers which have your number as their sum.

What are common difficulties which children encounter and how might these be overcome?

Addition with a zero – Children find it difficult to add when a zero is involved. They should realise that it means adding 'nothing'. When they have an answer of zero, they often need to be reminded to record it.

Adding in columns – When carrying out a calculation like $127 + 5 + 69$ some children confuse which numbers are units, which tens and which hundreds. The usual solution to this difficulty is for the children to write the numbers in columns.

Carrying – Some children have difficulty with a carried figure, for example:

$$\begin{array}{r} 57 \\ +26 \\ \hline 713 \\ \hline \end{array}$$

These children could carry out more examples using the Base Ten pieces and then linking each practical step to a recorded step.

SUBTRACTION

What does subtraction mean?

Subtraction, like addition, is a binary operation, that is, a procedure where two numbers are involved. The procedure can be presented in different forms:

- *taking away* – where the larger set is shown and a subset is removed leaving the 'answer' set, for example, 5 take away 2 leaves 3.

$$* \ * \ * \ \cancel{*} \ \cancel{*}$$

- *the difference between* – where both sets are shown and the answer set is shown by the unmatched members of the larger set, for example, the difference between 5 and 3 is 2.

$$\begin{array}{c} * \ * \ * \ * \ * \\ * \ * \ * \end{array}$$

- *counting on* – where the smaller set is shown and members are added to make up to the larger set, for example, 3 and 2 make 5.

$$* \ * \ * \ \ * \ *$$

  Subtraction is the inverse operation of addition, for example, $3 + 4 = 7$ so beginning with the sum and 'going back' to the first number involves subtraction, that is $7 - 4 = 3$.
  Subtraction is not commutative as the order of the numbers does matter, for example, $3 - 2 = 1$ but $2 - 3 = {}^-1$. Subtraction is not associative as the order of the operations

matters, for example, $(5 - 2) - 1 = 2$ but $5 - (2 - 1) = 4$.

What are real life examples of subtraction?

As a commodity is used up, the amount is subtracted or 'taken away', for example in eating, in spending money. When comparison takes place we can find 'the difference', for example in height, weight, area, volume, time, price. When giving change in a money transaction or finding how long an event will last we can 'count on'.

What is the key vocabulary and what does each word mean?

Subtract, take away, find the difference – all these mean the process of finding a quantity which when added to one of the two given numbers will result in the other. **Minus** is used to indicate that the second number is to be subtracted from the first and is recorded as $-$.

What is the historical background of subtraction?

Addition and subtraction were initially performed by means of a counting frame, and subtraction was linked to addition, for example 3 add 2 makes 5 so 5 take away 2 brings you back to 3.

The Egyptians who used a symbol of walking legs for addition, used the same symbol but 'walking' in the opposite direction, left to right, for subtraction. Stifel was the first European mathematician to use the signs + and − in a book on algebra. Signs like + and − are used to relate one number to another.

What is the value of teaching subtraction?

Subtraction is one of the basic operations which is carried out on numbers. Subtraction facts are used as part of the division process. As for addition, people subtract items, money, time and other measurements as part of their daily lives.

What are possible key steps in development for the learner?

1 The process of taking away involving 1 to 5
Children could learn the facts 'parcelled' as:
- 'take away 1' $(1-1, 2-1, 3-1, 4-1, 5-1)$,
- 'take away 2' $(2-2, 3-2, 4-2, 5-2)$
- 'take away stories' for 3 $(3-0, 3-1, 3-2, 3-3)$, 4 and 5.

The children could initially find the answers by putting out materials like Unifix blocks for the larger number, physically removing those to be 'taken away', then counting those left. Gradually, those 'left' will be recognised and counting will be unnecessary. Practice, possibly through games, should help the children remember the facts providing only a few are tackled at any one time.

2 'Take away' involving 0 to 10
Subtraction facts may be packaged in a similar way to those within 5.

3 Subtraction in the range of numbers 0 to 20
The subtraction facts may initially be presented

as 'take away', although steps 'back' on a number line might be used to find the answer. Then Base Ten pieces might be used to establish a more formal approach to the method known as *decomposition*. Here, for a subtraction such as $15 - 8$, the Base Ten pieces might be used on a notation card as follows:

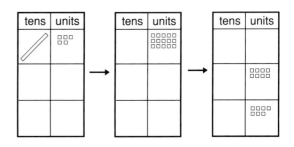

Language and written recording might be developed as:

$$^{0}\cancel{1}^{1}5$$
$$-\quad 8$$
$$\overline{\quad 7\quad}$$

Fifteen take away eight.
Fifteen is one ten and five units.
Subtract the units. I have five.
I want to take away eight. I can't.
Change one ten for ten units.
I have fifteen. I take away eight.
I have seven left.
There are no tens to subtract.
Fifteen subtract eight is seven.

4 Difference
This is initially introduced through materials like Unifix blocks where the two sets are displayed, matched in one-to-one correspondence and the number difference between them found. The more formal approach known as *equal additions* is based on finding the number difference but it is not widely used as a method in primary schools these days. In this method, the same amount is added to both numbers but in different forms as shown on the following page for $15 - 8$.

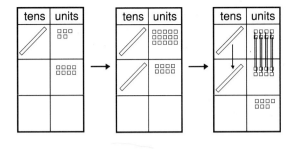

mentally. The procedure is to add on mentally in steps to the next ten, the next hundred etc. as shown in this example:

$234 - 87$

  87 and 3 are 90
  90 and 10 are 100, that's 13
 100 and 100 are 200, that's 113
 200 and 30 are 230, that's 143
 230 and 4 are 234, that's 147

 Complementary addition is the method used by shop assistants to give change, and by the customer to check change if it is dispensed by a machine, for example: A book costs £3·89 and the customer hands over a £5 banknote. The assistant would give change saying:

 £3·89 and   1p is £3·90
 £3·90 and 10p is £4
 £4 and £1 is £5
 The change is £1·11

The language and recording could go like this:

   Fifteen subtract eight.
$1^{1}5$ Subtract the units.
$-\ _{1}8$ How many more is five than eight?
 ———— It isn't. Add ten units to the top
 7 number.
 ———— Add one ten to the lower number.
   How many more is fifteen than eight.
   It is seven more.
   Subtract the tens.
   How many more is one ten than one ten?
   No more.
   Fifteen subtract eight is seven.

Regardless of how children are formally taught to carry out subtraction, most eventually develop their own form of mental counting on to do subtraction calculations.

7 Word problems
At all stages the children should meet subtractions presented as word problems and gradually recognise key vocabulary like 'left'.

5 Subtraction of tens and units
Base Ten materials are used for the decomposition method to carry out subtractions, then gradually children work the steps mentally.
 Zeros in a calculation often pose difficulties so children need plenty of practice with examples where zero appears as a digit in the larger number, in the smaller number and in the answer.

8 Decomposition with larger numbers
Base Ten pieces should be used to help children understand the exchange of one hundred to ten tens and later of one thousand to ten hundreds. For the children who become confident at exchanging, a 'short cut' could be introduced for numbers involving zeros, for example:

6 Subtraction by counting on
This method is known more formally as *complementary addition*. Examples using this method are most usually associated with money and time contexts. Children could learn initially by use of a number or time line and then

  ³⁹ When the child finds eight units
  4̶0̶¹3 cannot be taken from three units, it
 − 17 8 appears that there are no tens to
 ———— exchange to units. However, children
  22 5 with a grasp of place value will realise
 ————
there are forty tens and change one ten for ten units, leaving thirty-nine tens. The remainder of the calculation is carried out in the usual way.

9 Calculators

Calculators could be used for some subtraction calculations, patterns, and problems.

What are appropriate resources for subtraction experiences?

Unifix blocks, then Base Ten pieces are ideal to demonstrate the steps in a subtraction calculation. Taking 'backward' steps on a number line might be used. A number line can also be used for the 'counting on' method of subtraction. Coins are ideal for counting on when giving change. The analogue clock face or a time line can be used for counting on with minutes and hours. A calendar is ideal to count on in days.

What are possible contexts through which substraction experiences might be taught?

Shopping, TV and radio, journeys, the library etc. where duration times are calculated, are ideal contexts.

How might subtraction be assessed?

Assess *orally* by asking children to:
- explain what a subtraction calculation is
- answer mental subtraction calculations
- make up mental calculations to ask a friend
- talk their way through a formal written subtraction calculation as they do it.

Assess *practically* by asking children to:
- use Base Ten pieces to demonstrate to younger children how to do a subtraction.

Assess *in written form* by asking children to:
- carry out a graded selection of subtraction calculations
- answer about four subtraction calculations

- identify subtraction word problems from a selection of word problems
- make up a set of subtraction examples and produce a set of answers for them.

Assess through *problem solving* by asking children to:
- teach a young child to give change by adding on
- make a Bingo subtraction game.

What are common difficulties which children encounter and how might these be overcome?

The larger and smaller number – children should realise that in most subtractions (unless negative numbers are to be involved) the smaller number is subtracted from the larger. The children need to be aware there is a larger and a smaller number. However, this rule does not apply to digits which are only part of a number, for example, in the units column. In tens and units and other formal vertical subtraction calculations, children sometimes take the smaller unit number from the larger, regardless of whether the unit (or tens, or hundreds etc.) is the lower figure. This usually can be overcome by more practice using Base Ten materials and talking through the calculation.

Zeros – When faced with these within formal vertical calculations, many children find them confusing. They may require a greater understanding of the meaning of zero i.e. no units, or tens or hundreds.

Crutch figures – Some children carry out an exchange of a ten for ten units when this is not required and some forget they have carried out an exchange. Most children require to show an exchange with crutch figures. The children should be shown how these might be recorded neatly and clearly.

MULTIPLICATION

What do we mean by multiplication?

Multiplication is a short-cut method of carrying out addition where the sets have the same number of members, that is the same cardinality. Instead of adding $3 + 3 + 3 + 3$, we consider this as 'three added to itself four times' or 'four threes' and memorise the answer or 'product'.

Multiplication is a binary operation like addition and subtraction, that is two numbers are involved. Multiplication, like addition, is commutative and associative. For example, it does not matter if the multiplication 3×4 is calculated by multiplying the 3 by the 4, or the 4 by the 3, the order of the numbers does not matter, so the commutative law applies. The associative law holds because it does not matter which of two multiplication operations is carried out first, for example $3 \times 2 \times 5$ can be worked as $(3 \times 2) \times 5$ or as $3 \times (2 \times 5)$.

What are real life examples of multiplication?

If several of one item are bought, for example tickets for the theatre, tickets for a journey, cans of soup, packets of biscuits, the total price can be calculated by multiplication. Another application is to calculate the total number of seats in the aeroplane, the theatre, and the community hall by counting how many are in one row, counting how many rows have the same number of seats and then carrying out a multiplication.

What is the key vocabulary and what does each word mean?

Multiplier – the number of times another number is added to itself or the number by which another number is multiplied.

Multiplicand – the number which has to be added to itself or the number to be multiplied by the multiplier.

Product – the answer when one number, the multiplicand, is multiplied by another, the multiplier.

Multiple – when the number is an exact number of times another number. For example, 9 is a multiple of 3, 12 is a multiple of 2, 3, 4, 6 and 12.

What is the historical background of multiplication?

The multiplicative principle is older than the additive. It is believed to have been used in Egypt about 1600–2000 BC. The Egyptians indicated a multiplication by writing a smaller number before, below or above a larger number. In the Middle Ages, multiplication tables had not emerged but scholars did learn a 'twice times table' or the skill of doubling, and they used this to multiply by a method called duplation. Here is an example to show how it works:

$$23 \times 38 \text{ is interpreted as}$$
$$(1 + 2 + 4 + 16) \times 38 \text{ and then calculated as}$$
$$1 \times 38 = 38$$
$$2 \times 38 = 76 \text{ by doubling}$$
$$4 \times 38 = 152 \text{ by doubling}$$
$$\cancel{8 \times 38 = 304} \text{ by doubling}$$
$$16 \times 38 = 608 \text{ by doubling}$$

now cross out any lines not required and add to get

$$23 \times 38 = 874.$$

Scholars in the early seventeenth century did not memorise multiplication tables but they had rods which reproduced the tables. These Napier's rods are called after the Scottish mathematician. It is not certain if Napier was the first to use × as the symbol for multiplication, but it came into use about that time and is used in a translation of one of his books on logarithms.

What is the value of teaching multiplication?

It seems important that children should know about the operation because it exists. However, if children cannot remember their multiplication facts, they could make the calculation by repeated addition. Many learners adopt their own strategy to recall a product, for example, if the learner cannot remember the product of 7×9, he may remember 9×7. If this cannot be recalled either, the fact $7 \times 7 = 49$ may be remembered and then the learner adds 7×2, he knows seven sevens and adds seven twos. Another learner might remember 5×9 and then add 2×9, and so on. This mixture of multiplication and addition makes use of another law of arithmetic – the distributive law (the commutative and associative are others). It means that multiplication can be distributed over addition, for example:

$$7 \times 9 = 7 \times (7 + 2)$$
$$= (7 \times 7) + (7 \times 2)$$
$$= 49 + 14$$
$$= 63$$

or $\quad 7 \times 9 = (5 + 2) \times 9$
$$= (5 \times 9) + (2 \times 9)$$
$$= 45 + 18$$
$$= 63$$

Multiplication can also be distributed over subtraction, for example:

$$7 \times 9 = 7 \times (10 - 1)$$
$$= (7 \times 10) - (7 \times 1)$$
$$= 70 - 7$$
$$= 63$$

What are possible key steps in development for the learner?

1 Repeated addition patterns
Young children can use discrete materials such as animal shapes, toy cars, and Unifix blocks to set out patterns such as:

one three	***
two threes	*** ***
three threes	*** *** ***
four threes	*** *** *** *** and

so on.

They can also build up patterns like this:

three ones	* * *
three twos	** ** **
three threes	*** *** ***
three fours	**** **** **** and so on.

2 Learning a multiplication table
The patterns above show that tables can be built up in two ways:

- where the set has a specific number of members and another set with the same number of members is added each time, for example, one four, two fours and so on would be called the table of fours
- when the number of sets is specified and a member is added to each set each time, for example, four ones, four twos and so on would be called the four times table.

Most children learn the *times* tables. They learn the times tables for the numbers 2 to 10, some include the zero times and the one times tables. The majority, if not all children, learn the two times table first. The other tables may be tackled in any order, but most teachers have a preference. The order is often decided by relating one table to another, for example, the products of the two times table can be doubled to give the products of the four times table and these in turn can be doubled to give the products of the eight times table. The three, six and nine times table are often linked as are the five times and the ten times. The table which does not relate to any other is the seven times and the children will find some of the facts in this table the most difficult to learn by heart.

The steps to learn a table could be:

- build it with sets of objects or materials such as Unifix or Cuisenaire blocks
- build it by colouring squares to form columns with a common base level
- look at the pattern of the multiples, for example, for two, they are:

$$2, \quad 4, \quad 6, \quad 8, 10,$$
$$12, 14, 16, 18, 20,$$
$$22, 24, 26 \ldots$$

that is, all the unit numbers are even and in order. For three, they are:

$$3, \quad 6, \quad 9, 12, 15,$$
$$18, 21, 27, 30, 33,$$
$$36, 39 \ldots$$

that is, the unit numbers cover 0 to 9 but not in order

- group the facts for learning – the easiest are facts like 'times one', 'times two', 'times five', 'times ten' and the 'number times itself'; the most difficult are 'times six', 'times seven' and 'times eight'. Children could match groups of the facts, each written on a card, to the correct product, each written on a card
- realise that the order does not matter so that memorisation is greatly reduced
- play games to help memorise the facts – teacher's own games, like versions of Bingo, Dominoes and Snap can focus on a few facts at a time.

3 Multiplication of a single digit times two digits
The children could precede this work by considering an extension to their basic tables such as:

$$3 \times 1 = 3 \quad \text{so } 3 \times 1 \text{ ten} = 3 \text{ tens}$$
$$3 \times 2 = 6 \quad \text{so } 3 \times 2 \text{ tens} = 6 \text{ tens} \ldots$$

They would then attempt an example like

$$3 \times 23 \text{ as } 3 \times (2 \text{ tens} + 3 \text{ units})$$
$$(3 \times 2 \text{ tens}) + (3 \times 3)$$

or

$$
\begin{array}{r}
23 \\
\times 3 \\
\hline
9 \\
60 \\
\hline
69 \\
\hline
\end{array}
$$

Multiply the units.
Three times three is nine.
Multiply the tens.
Three times two tens is six tens.
Altogether there are six tens and nine units or sixty-nine.

Gradually this is carried out in one step, for example:

$$
\begin{array}{r}
37 \\
\times 3 \\
\hline
111 \\
\hline
_2 \\
\end{array}
$$

Examples, of course, should be graded so that the children learn the method and then learn to cope with carrying figures.

4 Multiplication with 10
The children could be challenged to look at the ten times table and develop a rule for multiplication with ten. Discourage 'add a zero' as this rule will fail them later for decimal fractions and could be interpreted as adding the zero to the left rather than the right. Wording like 'the units become tens' provides a model which can be extended to cope with all numbers.

5 Multiplication by numbers 11 to 19
The children should realise that they now combine the skills they have already learned (sections 3 and 4 above). For example: 13×26 is calculated as $(10 + 3) \times 26$

$$
\begin{array}{r}
26 \\
\times 13 \\
\hline
78 \text{ that is } 3 \times 26 \\
260 \text{ that is } 10 \times 26 \\
\hline
338 \\
\hline
\end{array}
$$

6 Multiplication by 20, 30, 40 . . .
Multiplication by 20, that is 10×2, is multiplying by 10 and then by 2. This is carried

out in one step by interpreting the multiplication by ten as 'calling the answer tens so recording a zero to the right', then the multiplication by two is done. The multipliers 30, 40, 50 . . . 90 are all carried out in one step in this way.

7 Multiplication by 21 to 99
Examples here combine multiplying by a multiple of ten and by a single digit.

8 Multiplication by larger numbers
The same stages, that is, multiplying by
- 100
- 100 to 199
- 200, 300 . . . 900
- 201 to 999
- 1000 and so on
should be tackled.

8 Word problems
The children should have experience of multiplication associated with real and simulated problems presented as word problems.

9 Using a calculator
Multiplication with larger numbers should make use of a calculator to save laborious effort. It is useful to ask some children to make a check on the correctness of their answer by approximating. For example, for 329×417, the answer is approximately 300×400, that is 120 000. This approximation is compared with the calculator display of 137 193 and the magnitude of the answer accepted as about right. A check may also be made by children carrying out the multiplication a second time and reversing the numbers, that is, making use of the commutative property of multiplication.

What are appropriate resources for teaching multiplication?

Cuisenaire rods or Unifix blocks are useful for building up tables. Squared paper, too, can be used to show columns of coloured squares.

Coins can be used to build up a table and this often is particularly interesting to children. For example, the four times table would be shown as:

four times 1p is 4p   (1p) (1p) (1p) (1p)

four times 2p is 8p   (2p) (2p) (2p) (2p)

four times 3p is 12p   (2p) (2p) (2p) (2p)
  (1p) (1p) (1p) (1p)

four times 4p is 16p   (2p) (2p) (2p) (2p)
  (2p) (2p) (2p) (2p)

four times 5p is 20p   (5p) (5p) (5p) (5p)
and so on.

Base Ten pieces may be used to demonstrate multiplication by 2, 3 and 4, but for greater multipliers, there are too many pieces to handle with ease.

Older children might like to produce a straight line graph for each table by relating the multiplicands on the x axis to the products on the y axis.

What are possible contexts through which multiplication might be taught?

The theatre, holidays, shopping, the supermarket and equipping the team.

How might multiplication be assessed?

Assess *orally* where pupils are asked to:
- answer randomly selected multiplication facts
- give several multiplication facts with a product of 24
- give another two numbers which have the same product as 4×9
- explain the steps as they multiply 23×24.

Assess *practically* where children are asked to:
- demonstrate with counters that 4×3 is the same as 3×4

- make a chart or a grid which could act as a 'times table reminder'.

Assess *in written form*, where children are asked to:

- calculate a set of multiplication calculations
- answer some multiplication word problems
- make up four multiplication calculations for a classmate to do
- make up two multiplication word problems
- make up ten multiplications for a classmate to do on a calculator.

Assess through *problem solving* by asking children to:

- make up a multiplication game where you write numbers on two cubes and use these as dice
- make up a multiplication track game.

What are common difficulties which children encounter and how might these be overcome?

Difficulty with recall of facts – Children should have the challenge of remembering a few facts, and then one table at a time. As each table is mastered, the child might be awarded a badge for the appropriate table. Some children may need to use a tables card as an *aide memoire* for a long time.

Difficulties with place value – Some children may find the rule for multiplication by ten, for example, 'call the units tens' confusing. Using Base Ten pieces to show that ten lots of six units become six tens sometimes helps. Multiplying numbers by 10 on the calculator might also help.

DIVISION

What does division mean?

The operation of division involves breaking down an amount or set into a number of smaller sets of the same cardinality, that is with the same number in each. There are two ways of doing this. One is to know how many should be in each set and then to find how many sets can be made. This method is known as *grouping*. The other is to know how many sets are required and to find how many are to be in each of these sets. This method is known as *equal sharing*. The same abstraction can be considered by either method, for example, $208 \div 8$ can be thought of as 'how many eights are there in 208?' (grouping) or as 'share 208 equally among 8 and find how many in each share?' (equal sharing).

Division is a binary operation as two numbers are involved. Division is not commutative as, for example:

$$4 \div 2 = 2 \text{ and } 2 \div 4 = \tfrac{1}{2}$$

that is, the order of the numbers matters.

Division is not associative as, for example:

$$(8 \div 4) \div 2 = 2 \div 2 = 1 \text{ and}$$
$$8 \div (4 \div 2) = 8 \div 2 = 4$$

that is, the order of the operations matters.

Division is the inverse operation of multiplication, for example:

$$3 \times 2 = 6 \quad \text{so} \quad 6 \div 2 = 3$$

What are real life examples of division?

The calculation is division where a total cost is shared equally, for example, a restaurant bill shared equally among a group of people. When a manufacturer wishes to know how many of an article he can make from an amount of stock, he carries out a division calculation, for example, finding the number of shirts which can be made from a bale of material.

When calculating the best buy, for example

among bars of chocolate, division calculations could establish the cost of 1 g of chocolate for each weight of bar to enable comparison.

What is the key vocabulary and what does each word mean?

Dividend – the amount to be divided.
Divisor – the number of sets or the number in a set.
Quotient – the answer to a division calculation.
Operation – addition, subtraction, multiplication and division are the four arithmetical operations.

What is the historical background of division?

The Greeks separated the dividend from the divisor by words while the Hindus often simply wrote the divisor beneath the dividend. In 1659 the Swiss mathematician Johann Rahn used the symbol ÷ for division in his book *Teutsche Algebra*. Before that, some writers had used the symbol to mean minus. In Europe the colon : was used for many years, until, due to a translation of the book *Teutsche Algebra*, the symbol ÷ was adopted in Great Britain and later in the United States.

What is the value of teaching division?

Division could be carried out by repeated subtraction. It seems sensible though, as this 'short-cut' to repeated subtraction exists, to show and explain the algorithm to children. The use of a calculator could be considered as a further short-cut.

What are possible key steps in development for the learner?

1 Practical informal activities
Young children carry out equal sharing tasks when distributing items among a group. Initially they are not thinking in terms of number but just follow through a process of 'one for you' until all the items are used up. They are very much aware of equal shares because they do not want less than anyone else. Later such tasks can be related to numbers, for example, 'If you share the 12 sweets equally among the 3 children, how many will each child get?' Grouping tasks like 'How many rows of 3 biscuits can you make on the baking tray with these 15 biscuits?' are also given to the children.

2 Subtraction and multiplication
Before children attempt formal division calculations they should understand the processes of subtraction and multiplication. It is beneficial if they know the multiplication table facts.

3 Division tables
It is useful for children to construct 'equal sharing' division tables from the 'times' multiplication tables. For example, a division by 3 table would begin:

$$3 \times 1 = 3 \quad \text{so} \quad 3 \div 3 = 1$$
$$3 \times 2 = 6 \quad \text{so} \quad 6 \div 3 = 2$$
$$3 \times 3 = 9 \quad \text{so} \quad 9 \div 3 = 3$$
$$3 \times 4 = 12 \quad \text{so} \quad 12 \div 3 = 4$$

The language used could be: 'three times one is three so three shared equally among three is one each' and so on.

It is also useful for children to consider remainders (shown here by the letter r) and the remainder pattern like this:

$$1 \div 3 = 0 \text{ r } 1$$
$$2 \div 3 = 0 \text{ r } 2$$
$$3 \div 3 = 1$$
$$4 \div 3 = 1 \text{ r } 1$$
$$5 \div 3 = 1 \text{ r } 2$$
$$6 \div 3 = 2 \text{ etc.}$$

Children could also consider how they might cope with the place value aspect of division by devising tables like these:

2 - 2

$$3 \text{ tens} \div 3 = 1 \text{ ten}$$
$$6 \text{ tens} \div 3 = 2 \text{ tens}$$
$$9 \text{ tens} \div 3 = 3 \text{ tens}$$
$$12 \text{ tens} \div 3 = 4 \text{ tens}$$

$$3 \text{ hundreds} \div 3 = 1 \text{ hundred}$$
$$6 \text{ hundreds} \div 3 = 2 \text{ hundreds}$$
$$9 \text{ hundreds} \div 3 = 3 \text{ hundreds}$$
$$12 \text{ hundreds} \div 3 = 4 \text{ hundreds}$$

and facts such as

$$7 \text{ tens} \div 3 = 2 \text{ tens r } 1 \text{ ten}$$
$$17 \text{ hundreds} \div 3 = 5 \text{ hundreds r } 2 \text{ hundreds}$$

4 Formal division

Base Ten pieces could be used practically to develop an algorithm based on equal sharing, for example:

14	*Tens*
4)56	5 tens to be shared among 4
−40	1 ten to each shares uses 4 tens
16	*Units*
−16	16 units to be shared among 4
	4 units to each share uses 16

This method can be used for division with any whole number divisor and where the dividend is either whole or decimal numbers, for example:

1·17	*Units*
24)28·08	28 units to be shared among 24
−24	1 unit to each share uses 24 units
4·08	*Tenths*
−2·4	40 tenths to be shared among 24
1·68	1 tenth to each share uses 24 tenths
−1·68	*Hundredths*
	168 hths to be shared among 24
	7 hths to each share uses 168 hths

5 Using a calculator

Using a calculator for division requires children to understand the meaning of the displayed answer, for example

$$82 \div 13 = 6.3076923$$

This needs to be interpreted as 'there are about 6 lots of 13 in 82' or '13 shares of 6 and there is a remainder'. The remainder is most easily found by a child carrying out a multiplication and then a subtraction like this:

$$13 \times 6 = 78$$
$$82 - 78 = 4 \quad \text{so} \quad 82 \div 13 = 6 \text{ r } 4$$

6 Meaningful answers

Often the answer to a division word problem depends on the context, for example:

- 82 peaches are packed in baskets of 13. How many baskets can be made up? The answer is 6 baskets.
- 82 people are to taste a new drink. A bottle holds enough for 13 people. How many bottles are required to give everyone a glass? The answer is 7 bottles.
- 82 litres of oil is to be poured into drums which each hold 13 litres. How many drums can be filled and what amount of oil is left over? The answer is 6 drums with 4 litres left over.

What are appropriate resources for teaching division?

Resources could include Base Ten pieces and money to illustrate the equal sharing process, games to help children learn the division facts, the calculator, and newspapers for data.

What are possible contexts through which division might be taught?

Class visits and events which involve equal sharing could provide division calculations. A baking project where the unit selling price is calculated, sports events and surveys where the average is found could also provide meaningful division work. Projects like holidays could involve a range of division calculations, for example, about costs, average temperature and speed.

How might division be assessed?

Assess *orally* by asking children to:
- describe each step in a calculation, like $425 \div 8$
- explain the meaning of an answer on the calculator, for example, $33 \div 7 = 4.7142857$
- estimate an answer for $200 \div 24$ and justify it.

Assess *practically* by asking children to:
- use Base Ten pieces to show the equal sharing process for a calculation like $56 \div 4$
- use string to find out how many 24 cm lengths can be cut from 2 m, and what length is left over.

Assess *through written examples* like:
- calculate
$$72 \div 3$$
$$426 \div 15$$
$$14 \cdot 6 \div 4$$
- find the edge length of a square whose perimeter is 90 cm.

Assess *through problem solving* situations such as:
- find the average amount of weekly pocket money received by the pupils in your group
- find the selling price for one of the biscuits you have baked so that you
  (*a*) cover the cost of the ingredients
  (*b*) make about 25% profit
- find the average speed for an athlete (a skier) or a car.

What are common difficulties which children encounter and how might these be overcome?

The processes involved – Some children need time to realise how multiplication and subtraction are part of a division calculation. The use of Base Ten pieces to illustrate the steps involved should help, especially if these are also linked to recorded steps.

Interpreting an answer – Children, understandably, find it difficult to realise an answer has different forms, for example, $21 \div 4 = 5 \text{ r } 1$ or $5\frac{1}{4}$ or 5.25. They also need to understand that the 'remainder of 1' is $\frac{1}{4}$ of another set of 4.

NUMBER PATTERNS

What does number pattern mean?

A number pattern is a set of numbers arranged in a definite order so that each number is connected to the next by a specific relationship or 'rule'. For example, to make a pattern:
- think of a rule, such as 'add three'
- decide on a number to begin the pattern, such as 20
- create the pattern 20, 23, 26, 29, 32, 35, 38 ...

A particular pattern may be chosen to represent numbers, for example, a pairs pattern would be built up as:

```
  *       for one        *  *   for two
*   *                     *  *
  *       for three      *  *   for four
```

and so on.

What are real life examples of number patterns?

The natural numbers are the simplest pattern: 1, 2, 3, 4, 5, 6, 7, 8, 9, 10, 11 ... where the rule is 'add one' and the pattern begins with 1.

Even numbers are also a pattern used

frequently in real life, for example, for numbering houses on one side of the street. Odd numbers would be used for the opposite side.

Each multiplication table gives a pattern of products. Patterns of dots allow numbers to be recognised on dice and dominoes, as do patterns of symbols on playing cards.

What is the key vocabulary and what does each word mean?

Series – another name for a number pattern.
Even number – a number exactly divisible by two.
Odd number – not an even number.

What is the historical background of number patterns?

Mathematicians have been fascinated by patterns in numbers from the beginning. For example, as the Egyptians recorded strokes, they did so in rows of three. The Chinese, too, recorded numbers as horizontal bars with a pattern layout. Pythagoras entertained large audiences by telling them about the qualities he applied to numbers and the patterns that this developed, for example, the first female number was 2 and the first male number 3, so 5, the union of these numbers, was for marriage. Great interest was taken in 'special numbers', for example, 'perfect' numbers are those where all the whole number factors add up to the number itself, that is, 6 is perfect as its factors add to give six: $1 + 2 + 3 = 6$.

There is interest in numbers where the dots can be arranged in a special shape, for example, 'triangular' numbers (as shown below) which are made by adding the next number in sequence:

$$
\begin{aligned}
1 &= 1 \\
1 + 2 &= 3 \\
1 + 2 + 3 &= 6 \\
1 + 2 + 3 + 4 &= 10
\end{aligned}
$$

Mathematicians developed other series which include square numbers (1, 4, 9, 16 . . .), pentagonal numbers (1, 5, 12, 22 . . .), and hexagonal numbers (1, 7, 9, 37 . . .).

A magic square where the rows, columns and diagonals add to the same total, like the one shown below, is believed to date back as early as 1000 BC.

4	9	2
3	5	7
8	1	6

Eratosthenes about 200 BC, discovered a way of placing whole numbers into two sets, those that are 'prime', that is, they are only exactly divisible by themselves and one, and those that are not.

John Napier's 'bones' which were multiplication tables used to simplify the multiplication of large numbers in the sixteenth century, were thought to be related to black magic.

What is the value of teaching number patterns?

Just like the mathematicians of the past, many children enjoy finding and creating number patterns. Such activities help them to understand more about the relationships between numbers and provide them with meaningful experiences in the four operations.

What are possible key steps in development for the learner?

1 The existence of pattern
As children learn about numbers they could investigate relationships among them, for example, adding two gives a pattern like 1, 3, 5, 7 . . . , subtracting two gives a pattern like 10, 8, 6, 4, 2, 0.

Ways of using objects, like animals going into the Ark, could lead children to see the even numbers, two, four, six, eight . . . as related to give a pattern. In the same way using an arrangement of pairs reveals numbers which don't produce pairs, like one, three, five . . .

2 Producing their own patterns
Children could try to make a sequence of patterns of counters, for example:

As children become familiar with numerals, they can invent their own patterns, for example:

$$
\begin{aligned}
1 &= 1 \\
1 + 1 &= 2 \\
1 + 1 + 1 &= 3 \\
1 + 1 + 1 + 1 &= 4
\end{aligned}
$$

They could also be asked to find patterns for given instructions, for example, 'What will happen if I begin with 1 and keep adding 4?' or 'What would happen if I start with 50 and keep subtracting 6?'

3 Multiplication tables
Children should investigate the products in a table as a pattern, for example, products in the five times table always end with 5 or 0. Such activities can also help children to see the patterns of multiples extending beyond the range of the numbers in a table.

4 Linking number and shape
Making shapes with counters and cubes can lead to triangular, square, hexagonal and other number series. Children could be encouraged to investigate other number patterns based on shapes, for example, the rectangle, the rhombus and the kite.

5 Continued patterns
Children can be challenged to continue patterns. Those where only two terms are given with the possibility of a range of different solutions could be particularly motivating to the

pupils as they try to find an unusual way of continuing the pattern. For example:

2, 8 might be continued as
2, 8, 14, 20 . . . (add 6 to give) or
2, 8, 32, 128 . . . (multiplied by 4 gives) or
2, 8, 512, 134 217 728 . . . (cubed gives).

Pupils can be asked to guess how friends have found the numbers they used, or asked to explain how their 'rule' makes a pattern.

6 Patterns with the calculator
Patterns are ideal to give pupils practice in a particular operation. For example, if they require practice in subtraction, provide patterns where they have to carry out lots of subtractions. Most calculations for patterns can be worked mentally, but more complex relationships or the use of larger numbers may lead to the use of the calculator. The use of the one key repeatedly to create the terms of a pattern is also appropriate, for example, in simple adding patterns such as $1 + 3 = 4$, the = can usually be pressed again and again to continue the 'add three' pattern.

7 Investigations
Children can be given investigations where they find or 'use' patterns, for example:
- look at the numbers 1, 2, 3, 4, 5, 6, 7, 8 and find a short way of calculating the total of these numbers ($1 + 8 = 9, 2 + 7 = 9 . . .$ altogether there are four pairs adding to make 9, so the total is four nines or 36)
- find how many diagonals you can draw in a quadrilateral, in a pentagon, in a hexagon . . . Predict from the number pattern of results how many diagonals you can draw in a decagon (2, 5, 9, 14 . . .)

What are appropriate resources for teaching number patterns?

Counters, cubes, a calculator and squared paper are useful but most of the investigations are likely to simply require paper and pencil.

What are possible contexts through which number patterns might be taught?

Numbers in history, patterns, the calendar, and linking shapes and numbers. If the children were developing their own magazine or newspaper, there could be a puzzle section where number patterns featured.

How might number patterns be assessed?

Assess *orally* by asking children to:
- say what the 'rule' is for simple patterns such as 1, 4, 7, 10, 13 . . .
- continue simple patterns such as 2, 4, 8, 16 . . .

Assess *practically* by asking children to:
- continue a pattern with pegs in a pegboard
- continue a pattern by colouring squares such as those shown below

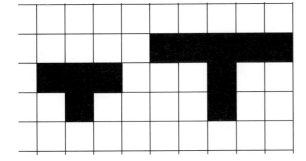

- select from a given set of number cards, for example, 2 to 32, to create a pattern.

Assess *in written form* by asking children to:
- create a pattern starting with a number such as 50
- describe the 'rule' used by a classmate to make his/her pattern.

Assess through *problem solving* by asking children to:
- complete this number pattern based on addition using the square numbers; it begins:

$$1 = 1$$
$$4 = 1 + 3$$
$$9 = 1 + 3 + ?$$

What are common difficulties which children encounter and how might these be overcome?

Creating and identifying patterns – Some children may find it easier to create patterns than to identify the relationships in given patterns. Others may prefer the challenge of finding relationships and continuing patterns before they have the confidence to create their own.

AVERAGES

What does average mean?

The arithmetic average or 'mean' of a set of numbers or quantities is calculated by dividing their sum by the number of members. For example, if five pieces of cheese have the weights 250 g, 250 g, 260 g, 270 g and 280 g, the average weight of cheese is calculated in the following way:

Total weight is
250 g + 250 g + 260 g + 270 g + 280 g = 1310 g

Average weight is 1310 g ÷ 5 = 262 g

The calculation illustrates the *mean* as a theoretical form of equal sharing to give one quantity which might be used as a 'representative' of all the quantities in a set.

There are two other terms called the *mode* and the *median* which are also forms of averages. The mode is the number or quantity which appears most frequently in a set of numbers or quantities. For example, for the pieces of cheese listed, the mode weight is 250 g, because two pieces have that weight and there is only one piece of each of the other weights. The median is the middle value in a set of numbers or quantities arranged in order of magnitude. For the cheese pieces, the median weight is 260 g as there are two weights less than this and two weights greater than this. If there is an even number of quantities, the median is the average of the two middle terms.

What are real life examples of averages?

The sizing of clothing and footware is based on average measurements so that most people can find a size that is suitable for them. If most people are suited, it is likely that the mode measurements are the average used.

Football teams can be classified by their goal average. Cricket players can have a batting and bowling average. Here the scores are the mean values.

In a race, children might hope to run it in less than the average time. However, it is likely that they will be pleased if they finish the race among the faster half of the runners, that is that their time is less than that of the median value.

The mean speed is **not** calculated by finding the sum of the miles or kilometres per hour data and dividing. For example, 30 mph and 50 mph does not have an average speed of 40 mph – this requires a different form of calculation where time is considered.

What is the key vocabulary and what does each word mean?

Average – can mean ordinary or usual; mathematically there are the three meanings already given.

Range – a set of values is usually spread in magnitude from a lowest value to a highest value; numerically the range is expressed as the difference between the highest and the lowest values.

What is the historical background of averages?

The word average first appeared about 1500 in connection with Mediterranean sea trade. Increasing attention to psychology, sociology and economics in this century have greatly increased the use of averages.

What is the value of teaching averages?

In every day life, the average is useful when discussing what could be called social statistics such as earnings, diet, personal measurements.

What are possible key steps in development for the learner?

1 The mode
This is probably the easiest concept of average for young children. For example, their clothes are sized according to the mode for the age group, their books are written with vocabulary which most of the age group can read and the food most of them like is bought by shopkeepers to sell to them. In the classroom, surveys can be carried out to find the mode value for the class of, for example, height, weight, amount of pocket money, number of siblings, favourite TV programme, favourite pop star and favourite football team. The children can decide what they want to collect data about, make out a table on which to record responses and then compare the tally marks to find the most common response, that is, the mode.

2 The median
The median could be the next concept of average which pupils learn about. The children decide what they want to collect data about and form the question to ask classmates. Responses should be organised according to magnitude and then the median identified, or where necessary calculated as the mid-value of two terms. The children might want to use and

extend some of the data they have collected for the class to produce display figures called Sara Median and Sean Median. Such figures, if life size, could be used for comparison of height and length of limbs.

3 The range

It is useful to consider the range of collected data. How near is everyone to the mode or to the median value? The children could be asked to carry out a task like writing the name of the school or completing a worksheet of calculations within a specified time such as one minute. The results could be recorded and the lowest and highest number (of the written name or correct calculations) noted. The children might want to know whether the spread of the number of times the school name was written is greater than the spread for correct calculations. They can be shown how to find the range by subtracting the lowest value from the highest.

4 The mean

Many people only interpret the average as the mean value and the emphasis in primary school can be mainly on calculating the mean. It seems essential that any calculations are related to a context which gives understanding about the mean and a purpose for calculating it.

The mean can also be explained through a study of the physical build of children. Each group could select a physical characteristic like height, weight, waist size, shoe size, shirt size or sock size. Data could be collected from each pupil in a group and the results recorded as a bar graph. It should be obvious that some children have a greater measurement and some a smaller measurement. The children could reason where they might draw a line that represents an 'equal share' value, like the one shown in the diagram opposite.

5 Calculation of the average or mean

The children could consider how to calculate the average having discussed what the average is and having had a visual representation of the

average for some sets of measurements. They might decide to find the average by:
- making all the bars of a bar graph the same height, adding to some and subtracting from others, or
- realising that all the individual measures must be taken into account, adding them and finding the average by division, using the number of measurements as the divisor.

The result may not be one of the numbers or quantities in the original set and some children may need an explanation that the equal sharing which gives the average is not a 'practical' process. Sometimes the result will not be a whole number and pupils could discuss what accuracy of answer is the most meaningful. For example:
- What is the average amount of pocket money for the following group of children where Ben receives £2·80, Fiona gets £3·50, and Jill gets £5?
  Total money is £11·30. Number of children is 3. Average amount of pocket money is £11·30 ÷ 3 = £3·76666 . . . or £3·77 to the nearest penny.
- What is the average number of strokes per hole taken by a golfer with the following scores: 3, 4, 5, 5, 2, 3, 6, 3, 2?
  Total strokes is 33. Number of holes is 9. Average strokes per hole is 33 ÷ 9 = 3·66666 . . . or 4 to the nearest stroke.

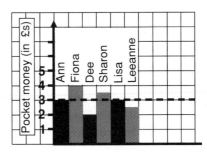

6 Investigating mean values

The children could find the mean of events which interest them, for example, for the pupils in a group:
- How many pages can be read in ten minutes?

- How long does it take to look up and write meanings for ten words?
- What height can be jumped over with two feet kept together?
- How long does it take to do twenty-five press-ups?

7 *More complex calculations involving the mean*
A few pupils may like the challenge to use given data to identify:

- a missing number/quantity when the others and the average are given
- how many numbers/quantities are in the set if the total and the average are given.

What are appropriate resources for teaching averages?

Prepared data (selected to give the children exactly divisible totals initially), data about themselves, the school, the locality, games and sports, squared paper to draw graphs, calculators.

What are possible contexts through which averages could be taught?

Ourselves, families, health, sport, and many other contexts where information about average likes, dislikes, needs and sizes are involved.

How might averages be assessed?

Assess *orally* by asking children to:

- explain how to find the mean of the following numbers: 2, 3, 5, and 6
- describe in their own words what the mode value is.

Assess *practically* by asking children to:

- find the mean number of flowers on a set of plants
- find the mode for the number of sweets in several bags
- find the mode for the number of pens/pencils which belong to a group of pupils

- find the median number of 'skips' with skipping ropes for both the boys and the girls in the class.

Assess through *written examples* such as:

- To enter a race William must have an average time in the trials of $3\frac{1}{2}$ minutes. So far he has times of 3 min 50 sec and 3 min 25 sec. What time, or less, must he get for the third and last trial to gain entry?

Assess through *problem solving* situations such as:

- You are going to produce a sweat shirt which will fit most of the boys in your class. Make a paper pattern for it.
- Make up a knock-out competition where it is the player's average score which takes him/her into the next round.
- Find out which generation seems to have taller men – Dads or Grandads.

What are common difficulties which children encounter and how might these be overcome?

Realising that the mean often does not exist – When children use real data and understand that some of it will be above the mean and some below, they expect the mean to be one of the measures rather than a different measure. This difficulty might be resolved by using, for example, five plastic bags containing different weights of pasta shells – 200 g, 250 g, 250 g, 300 g, 400 g. The bags should be labelled with the weights. The total weight of pasta shells is found and this is separated out equally into another set of five bags. Each of these bags will be labelled with 280 g, the mean value. The children can see that this is different from the amount held by any of the original bags.
Interpreting results which are not whole numbers – If the data comprises whole numbers, some children might find it difficult to know how to express a decimal answer. Such examples need to follow work on rounding decimal fractions and expressing answers to a given degree of accuracy.

PROBABILITY

What does probability mean?

Probability is a measure, or estimate of the likelihood of an event happening. It can be expressed in qualitative terms such as 'very likely', 'certain', 'equal chance', 'impossible' or on a numerical scale which extends from 0 to 1 with 0 as impossible.

What are real life examples of probability?

The example which must spring to most people's minds is the weather forecast where the expert is involved in predicting what will happen. Insurance companies also try to predict how long clients will live when they offer terms on life insurance. It is the trend of the weather and the proportion of the population who die at a particular age which is predicted, not the actual weather or the actual age of death of the person. Players of many games, in which chance is an element, like bridge, require to become expert in predicting what their opponents can and will do.

What is the key vocabulary and what does each word mean?

Probability or likelihood – can also be calculated as the proportion of favourable outcomes to the total number of possible outcomes.
Outcome – possible result.
Random – haphazard, chosen without any regard to any characteristics of the individual items.

Qualitative terms explain themselves, for example, if the probability is *'impossible'*, it will not happen and this can be given the numerical value of 0, whereas *'certain'* means it will happen and the numerical value is 1.

What is the historical background of probability?

About 1650 a gambling problem was sent to the French mathematician Pascal. He asked a colleague, Fermat, to work it out too. They both found the same answer but used different methods and as a result of a following discussion the idea of mathematical probability was conceived.

A Swiss mathematician, Bernoulli, later in the same century, also contributed to the theory of probability. Early in the eighteenth century, De Moivre, a French mathematician, published ideas about probability. As an old man, he fell into poverty and supported himself by solving questions about games of chance.

What is the value of teaching probability?

The increasing importance of psychology, sociology, statistics and economics has resulted in a greater emphasis being placed on mathematical probability.

What are possible key steps in development for the learner?

1 The concept and language
A good beginning is to consider events, especially those in the children's lives, which are 'certain' and 'impossible'. For example, the children can find real events which are impossible, such as 'baby brother coming to school tomorrow and doing my work for me', and can invent variations of real life such as 'a boy with a tail', and 'a cat with two tails'. They can also devise events which are thought to be certain such as 'I'll need to eat to stay alive' and 'the sun will rise tomorrow'. During discussion about events, some will be mentioned that

cannot be considered as impossible or certain, and other terms will come into use to describe them, for example, 'a good chance', 'a poor chance', 'very likely', and 'very unlikely'.

2 Ordering events

The children could be provided with a set of events which they are to place in order from impossible to certain. They could also consider terms for the probability of these events and order these like this:

> impossible,
> very unlikely,
> unlikely,
> equally likely,
> likely,
> very likely,
> certain

The meaning of 'equally likely' or 'an even chance' would require discussion to conclude that these terms are used for events which are just as likely to happen as they are not to happen. The terms 'fair' and 'unfair' would require discussion. The children should be able to suggest situations they regard as unfair, for example, if there was a raffle and some tickets were left out of the draw. This would mean it was impossible for some tickets to be prize winners. The children might consider the 'fair' criteria for a raffle to include:

- all the tickets put in the draw
- all the tickets very well mixed around.

3 Outcomes

The children should discuss what could be possible outcomes both for real and theoretical situations. For example, if a cube is hidden under one of two paper cups, possible outcomes for the first cup are:

- the cube is there
- the cube is not there.

If the cube is under the first cup, then there is only one outcome for the second cup – that the cube is not there; if the cube is not under the first cup then, again, there is only one outcome for the second cup.

4 A numerical scale

The children should find it easy to accept that if an event is impossible it is regarded as having a probability value of 0 as it has a zero chance of happening. It is more difficult to explain that if an event is certain, it has a probability value of 1. Again, it is easier to accept that an equal chance has a numerical value of $\frac{1}{2}$ as it is 'in the middle', 'halfway between' or 'will be one out of two possibilities' (certain or impossible).

5 Probability values

The children could extend the notion of probability having a numerical value by investigating the number of possible outcomes for events. For example, when asked whether they have blue eyes and brown hair, it is possible to have both, it is possible to have neither, and it is possible to have blue eyes but not brown hair or brown hair but not blue eyes – there are four possible outcomes. If a class of twenty-four children carried out this survey, it could be predicted that one out of four, that is six, would have both blue eyes and brown hair, six would have neither, six would have only blue eyes and six would have only brown hair. However, it is unlikely that the class would have exactly that composition. Children should realise that in any survey the prediction is just an estimate because only a sample of the population is involved.

What are appropriate resources for teaching probability?

Coins to toss, dice to throw, playing cards to select, beads of different colours and a bag to hide them in, games to discuss for outcomes, fairness and probabilities.

What are possible contexts through which probability might be taught?

Games, the weather and ourselves.

How might probability be assessed?

Assess *orally* by asking children to:

- discuss the probability of a range of events, for example, the headteacher coming into their classroom before lunch, and there being baked potatoes on the school menu
- tell you the meaning in their own words of terms such as 'a fair situation', 'an even chance' and 'certain'.

Assess *practically* by asking children to:

- identify what outcomes are possible, what the probability of each outcome is and what actually happens when you have coloured beads in a bag, for example, two blue and four red.

Assess through *written examples* like:

- using the four king cards, face down, from a pack of playing cards, list possible outcomes when one card is selected.

Assess through *problem solving* by:

- devising a 'magic' trick using playing cards where you can select a card and be very likely to predict something about it.

What are common difficulties which children encounter and how might these be overcome?

Concepts – Many of the concepts are sophisticated for young children to understand. However, if examples are kept practical and the terms used discussed with the children so that they develop an understanding of them, they could enjoy this aspect of mathematics.

RATIO

What does ratio mean?

A ratio expresses the relationship between two numbers or quantities. The two quantities are compared by being expressed in the same unit so that the ratio can be thought of as a number without reference to any particular units. For example, for the scale on the map where 1 cm represents 1 m or 100 cm, the ratio would be expressed as one to a hundred and written as 1 : 100. Another example of a ratio could be the comparison of two lines, one of which measures 10 cm and another which measures 20 cm. The ratio would be expressed as ten to twenty (10 : 20) or one to two (1 : 2) as ratios can be simplified just like common fractions to their simplest form. In a ratio, one part is compared with another part, whereas the fraction is expressing the relationship between one part and the whole. For example, if a sum of money is divided between two brothers so that the elder gets £5 for every £2 the younger gets, the ratio is 5 : 2 whereas the fraction the elder gets is $\frac{5}{7}$ and the fraction the younger gets is $\frac{2}{7}$ of the total amount.

Many people will associate ratio with trigonometry and remember that the sine of an angle in a right-angled triangle is the ratio of the length of the opposite side to the length of the hypotenuse.

A ratio arises from the comparison of two quantities in the same unit. A comparison of two unlike quantities gives rise to a rate. Sometimes these two terms are confused. Some examples of rate are:

- typing, expressed as words per minute
- bricklaying, expressed as bricks per minute
- speed, expressed as miles/kilometres per hour.

What are real life examples of ratio?

The gradient or slope of a hill is measured as a ratio. A ratio of 1 : 10 is interpreted as a rise of 1 metre for every 10 metres forward.

What is the key vocabulary and what does each word mean?

Ratio – a relationship between two numbers or quantities which is without dimensions.
Rate – a relationship between two linked but different quantities.
Proportion – a relationship among four numbers or quantities in which the ratio of the first pair is the same as the ratio of the second pair.
Similar – where two 2D shapes have corresponding angles equal and so all corresponding edges are in proportion. This results in one shape being an enlargement of the other.

What is the historical background of ratio?

The Greeks saw ratios of lengths as important in buildings. They conceived the 'golden ratio', that is the ratio which divides a line so that the ratio of the longer length to the shorter length is the same as the ratio of the whole length to the longer length.

The notation used today is the colon, for example, $2:3$. This first made its appearance in Britain in 1651 in a publication by an astronomer, Vincent Wing.

What is the value of teaching ratio?

It is another form of expressing comparisons. It is involved in exchange rates and recipes.

What are possible key steps in development for the learner?

1 Rate
This is possibly an easier idea than the more abstract ratio and could introduce pupils to the comparison mechanism. For example, the pupils could calculate their own rates for:
• writing words per minute

• buttoning buttons per minute
• toe-touching per minute
• press-ups per minute, and so on.

2 Ratio
This might be introduced as quantities for a party drink. For example, one litre of a fruit cocktail could contain: 100 ml passion fruit juice, 300 ml pineapple juice, 200 ml orange juice, and 400 ml sparkling mineral water. The ratio of passion fruit juice to pineapple juice is $100:300$, and the ratio of orange juice to water is $200:400$. If the quantities are to be increased, they could be doubled, trebled etc. but they must still be in the same ratio. This would also involve the children in realising that, for example, $200:400$ is the same as $2:4$ and $1:2$.

3 Finding missing dimensions/quantities using ratio
'If 100 ml of passion fruit juice is used with 200 ml of orange juice, then the ratio of passion fruit juice to orange juice is $1:2$. If the person mixing the drink starts with 150 ml of passion fruit juice, what is the amount of orange juice to be used?'
Children should realise that when attempting a word problem like this they are involved in multiplication and division as they find the ratios.

What are appropriate resources for teaching ratio?

Recipes, card/plastic coins, maps and data of all kinds, calculators.

What are possible contexts through which ratio might be taught?

A project on a country like Australia; a journey, ourselves.

How might ratio be assessed?

Assess *orally* by asking children to:
• explain what ratio is in their own words

- find the missing length by looking at each of the lines below which both have two parts in the same ratio

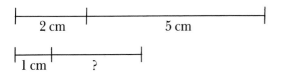

2 cm   5 cm

1 cm   ?

Assess *practically* by asking children to:
- lay out fifty cubes in the ratio 2 : 3
- make a similar triangle with geo-strips where the ratio of the sides is 1 : 2.

Assess *in written form* by asking children to:
- rewrite this recipe for Corned Beef Hash which is for four people so that it can be used for six people

    50 g margarine
    1 chopped onion
    350 g diced corned beef
    450 g diced cooked potatoes

    2 cooking apples
    1 tablespoon Worcestershire sauce
    seasoning to taste
- draw two triangles in the ratio 1 : 5.

Assess by *problem solving* by asking children to:
- invent a question for classmates where they use their knowledge of ratio.

What are common difficulties which children encounter and how might these be overcome?

The concept – Many children find it difficult to grasp what ratio is about. Carrying out practical exercises where items are laid out in a specified ratio might help.

Multiplicative – It is important to emphasise that doubling, trebling, halving, quartering – that is multiplying and dividing – are all means of finding different quantities in the same ratio. Some children try to add.

ALGEBRA – EQUATIONS

What does algebra mean?

Algebra is a language through which mathematicians express relationships. They use numbers and symbols like a shorthand version of words. For example, 'three add two gives five' would be written as $3 + 2 = 5$. To understand the written symbols of algebra, the learner should think in terms of the words. However, the symbols and the underlying words become more difficult when the mathematician uses letters to represent a range of numbers rather than a specific number. For example: 'two multiplied by itself is called two squared' and is written as $2 \times 2 = 2^2$, but $a \times a = a^2$ meaning 'any number times itself can be written as the number squared' is much more difficult to express in words and to understand. Writing numerals and symbols is not only more succinct than words but is particularly useful for expressing rules that always apply.

What are real life examples of algebra?

Situations where conditions apply can be expressed in algebraic terms, for example, hiring a car where there is a basic charge and then a mileage charge, and paying the phone bill where there is a fixed rental charge and a charge per call which varies according to time and distance. You can write a formula for expressing the way to change

- a temperature in Fahrenheit to one in Centigrade
- an amount in £s to an amount in another currency
- a time in London to a time in another city.

What is the key vocabulary for algebra and what does each word mean?

Variable – a symbol, such as x or any letter, used to represent an unknown number, quantity or a set of values.

Formula – a sequence of symbols used to express a rule or a result.

Equation – a formula where two expressions have the same value.

Expression – any symbol or string of symbols.

What is the historical background of algebra?

The Egyptians and the Greeks were concerned with expressing words as numbers, and conceived problems with 'unknown' numbers. However, it was the Arab mathematician al-Khwarismi who, in the ninth century, wrote a book with a title in which the word 'al jabr' appeared that led us to the word 'algebra' and the use of symbols as representations of words.

What is the value of teaching algebra?

Mathematicians like to write everything as succinctly as possible. If children learn to do this themselves much of the mystery that surrounds symbols could disappear.

What are possible key steps in development for the learner?

1 Using symbols

Children should be given the task of writing something as symbols rather than words, for example, 'two add four makes six' as $2 + 4 = 6$. They could also interpret signs as words, for

example, $24 \div 4 = 6$ as 'twenty-four shared equally among four gives six to each' or 'twenty-four put into fours gives six lots'.

2 The notion of algebra

The learners could think of algebra as a convenient way of discussing number without using any particular numbers. A possible beginning could be to discuss a 'guess the number game' and show how a solution might be expressed in numbers and signs. For example, 'I am thinking about a number. If I add two, the answer is five', could be expressed as $x + 2 = 5$. Another example might be 'What do I add to 4 to make 9?' where the children would record $4 + x = 9$. Both of these examples show how a letter can be used to represent an unknown number. The children can use their own choice of letter, but should know that x is often used by mathematicians. As the pupils grow in confidence with the use of algebraic statements, these can become more complex, for example, 'I am thinking about a number and when I multiply it by itself and add one, the answer is thirty-seven' could be written as $x^2 + 1 = 37$. Initially such statements will be solved by trial and error rather than by any formal approach.

3 Equations

Cubes, coins, and other objects could be used to illustrate the algebraic statement. If, for example, the pupils are trying to find $5 + x = 8$, they might place 5 cubes on one pan of a balance and 8 on the other. The task is to find how many more cubes would be added to the pan with 5 to make a balance. This concept of an equation having a balance is important to the pupils as a foundation for later work.

4 Examples where there is more than one solution

The next step might be to consider an example which allows a range of solutions – 'I am thinking about a number which when I add one gives an even number'. This is more difficult to express in words but the use of a second variable could help, that is, $x + 1 = y$ where y is any even whole number. Here x, representing the number being 'thought about', could be given each of a range of values and y calculated each time, that is:

If x = 1 then 1 + 1 = 2 which is even, so x could be 1.
If x = 2 then 2 + 1 = 3 which is odd, so x can't be 2.
If x = 3 then 3 + 1 = 4 which is even, so x could be 3.

The pupils may realise that x can be any odd number and that they have 'proved' this.

What are appropriate resources for teaching algebra?

Cubes, coins, marbles and a balance can be useful.

What are possible contexts through which algebra might be taught?

Games, ourselves, and costs.

How might algebra be assessed?

Assess *orally* by asking children to:
- explain in words what each of these could mean:

$$1 \times b = a$$
$$c = 2\pi d$$
$$x + 4 = 7$$
$$2x = 8$$

Assess *practically* by asking children to:
- use the balance and cubes to find x if $3 + x = 7$.

Assess *in written form* by asking children to:
- find x if $x + 4 = 11$
- write the following in numbers, letters and signs: 'Andrew is three years older than William.'

Assess through *problem solving* by asking children to:
- write word sentences then equations (number and letter sentences) about
  - how to change any metre measurement to a centimetre measurement
  - how to change any time expressed as pm to 24 hour notation.

What are common difficulties which children encounter and how might these be overcome?

Equations – Some children find it difficult to accept the use of letters for unknown numbers. They are likely to require a great deal of practice expressing statements in words before thinking of these as numbers, signs and letters.

MONEY

What does money mean?

Money in most countries is produced as metal coins and paper banknotes. These are used for payment, either to purchase goods or to pay for a service. Today the use of credit cards has eliminated the use of actual money in many transactions. Internationally, the purchasing power of money is determined by the confidence placed in the financial stability of the government who issues it.

What are real life examples of money?

In the United Kingdom seven coins are in general use at present. These have the following values: 1 penny, 2 pence, 5 pence, 10 pence,

20 pence, 50 pence and 1 pound. Each coin can be identified by its shape, colour, size and design, as well as by the value which appears in numerals, and/or words. Banknotes for the values £1, £5, £10, £20, £50 and £100 are rectangular in shape and vary in size and design. The £1 note is no longer issued by English banks but continues to be produced by Scottish banks.

What is the key vocabulary for money and what does each word mean?

British currency: **penny** (p) – basic unit
pound (£) – 100 pennies or pence.

The pence coins, despite their present metal composition, are referred to as 'copper' and 'silver'. Other words include:

Price or cost – the amount of money for which an item is bought or sold.
Profit – basically the difference between the buying and the selling prices.
Discount – an amount of money taken off the selling price.
Expenses – an amount of money used in the performance of a task.
Mint – the place where money is produced.
Bank – an institution for the keeping, lending and exchanging of money; a container for keeping money as savings.
To buy – to exchange money for goods or services.
To spend – to use money for buying.
To save – to avoid spending an amount of money or to set money aside for future use.

What is the historical background of money?

The word 'money' is derived from the Latin word *moneta* which is the name given to a Roman goddess. The temple built in her honour was also the site of Rome's first mint.

In the course of history money has taken many forms such as shells, tusks, wooden rods, feathers, cattle and salt. The first known coin was used in Asia Minor in the seventh century BC. This developed from metal bars or strips which could be broken or weighed. The introduction of the coin proved to be a boon to trade and spread rapidly because of its convenience. 'Money alone sets all the world in motion', said Publilius Syrus, a Roman who lived in the first century BC. This view is held by many today.

What is the value of teaching money?

Handling money is a necessary skill for life. It is essential that all of us are able to:
- recognise the value of each coin and banknote in our own currency
- make up the same amount with different coins and/or notes
- interpret and record an amount of money
- match recorded amounts to the correct coins and notes.

What are possible key steps in development for the learner?

1 The concept of money
Children come to school with different experiences and this is particularly true about money. Some will have used money in a shop, with and without parental guidance, while others may never have handled money. Children should realise that coins may be exchanged for goods and this might be developed through a class shop where the children use one pence coins merely to exchange as a token for an item. Vocabulary for the children to understand and to learn to use might include 'buy', 'shop', and 'money'.

2 Language and the concept of value
The class shop can be used as an application of

early number work and amounts of pence may be counted out to pay for slices of fruit and biscuits. The children should now be focusing on the concept of value in that items are exchanged for different amounts. The children should look at the penny coin so that they can describe its colour, shape and design. They should be able to recognise the penny coin among other coins. Vocabulary could include 'coins', 'penny', 'pence', and 'pay'. Some children may not know how to count pennies and think of them as 'one penny, one penny, one penny . . .' They should develop language such as 'one penny, two pence, three pence . . .'

3 One coin for a number of pence

Some young children find it a difficult step to see one coin and name it as an amount like two pence or five pence. They usually need practice in showing two as two discrete cubes *and* as one tower of two before being introduced to the two pence coin. The coin has the additional difficulty in that it does not look like two one pennies. This is an extension of the concept of value. Some children find it easier to accept the five pence coin as equivalent to five pennies, possibly because of its difference in colour, and it may be sensible to introduce this before the two pence coin.

4 Coins and amounts to 10p

Children should be able to:

- recognise the 1p, 2p, 5p and 10p coins
- be able to place the coins in order of value
- relate the 2p, 5p and 10p coins to other coins
- show the amounts 2p to 10p using different coins
- count sets of mixed coins, usually by initially placing these in order of value.

This learning should be developed using coins, especially in a shopping context. Most transactions should involve the purchase of one item with the customer noting or asking the price, counting out the required amount and handing it to the shopkeeper who checks it. To reinforce addition within ten, purchase of two items might be included. The child can add then find the correct coins, or find the coins for each item and then add the coins.

5 Coins and amounts to 20p

The use of the 10p coin with pennies is useful to follow work with Base Ten pieces to help the children in their understanding of place value. The 20p coin is introduced. A start might be made to subtraction as a counting on process for giving change.

6 Coins and amounts to 50p

The children could reinforce multiplication by counting in twos using 2p coins, in fives using 5p coins, in tens using 10p coins. Mixed sets of coins for amounts 21p to 50p should be ordered with the largest value first and then counted. Addition and subtraction (e.g. finding the difference in price between two items) experiences and word problems could be carried out. The 50p coin should be introduced and related to the other coins.

7 Coins and amounts to £1

The £1 coin (and note in Scotland) should be introduced, if this has not been done earlier. Because of the difficulty of making early work in number related to real life, the £ might be introduced alongside the penny. It could be introduced as the coin used to buy dear or expensive things. The relationship between the coins could be left until this stage in the development.

Addition, subtraction, multiplication and division money examples can be included with those in number. Shopping activities, possibly related to measure work, for example, through a greengrocer's shop where items are priced by weight, could be used. Shopping and other number games are useful for experiences of exchanging coins.

8 Amounts over £1

The notation initially may be introduced with

the 'point' explained as the separator between pounds and pence and later as the marker of the units in decimal notation. The banknotes should be introduced and simulated copies used in shopping transactions and games. Examples involving money could include addition, subtraction, multiplication and division as well as averages, simple common fractions, decimal fractions and percentages. Some of these examples may be calculated mentally while others involve an approximate mental calculation and an exact answer using a calculator.

9 Applications of money
Pupils could apply their knowledge of money to real situations. For example, the cost of school lunches for a term, the cost of a mountain bike through hire purchase, profit and loss transactions in the school tuck shop could be discussed and calculated, the cost of a holiday abroad and the foreign currency used could be investigated.

What are appropriate resources for teaching money?

Real, cardboard, gummed paper and plastic coins, items for the stock of a shop, a 'till', price lists, menus, catalogues, order forms, receipts, purchases print-outs, and 'for sale' advertisements are just a few of the possible resources.

What are possible contexts through which money might be taught?

Visits could be made to, or classroom simulations set up for a variety of shops, for example, the fishmonger, the fruiterer, the florist, the baker, the café, the post office, the supermarket; mail order catalogues; the travel agent using holiday brochures.

Opportunities should be taken for using real money in the organisation of school events like the tuck shop, the school charity collection, the bring and buy sale, the book club, and the concert or show.

Contexts such as shopping, the supermarket, holidays, decorating my room, and many others can provide experiences in money.

How might money be assessed?

Assess *orally* by asking children to:
- identify a coin
- describe an unseen coin for classmates to guess
- describe the similarities and differences between two coins
- count out the sum of some mixed coins
- carry out mental calculations.

Assess *practically* by asking children to:
- set out a given amount
- show a stated amount using two different sets of coins
- find the fewest coins to show a stated amount
- give change in a shopping transaction
- use a calculator for money calculations.

Assess *in written form* by asking children to:
- write labels for the prices of shop items
- write a receipt for a shopping transaction
- carry out calculations.

Assess through *problem solving* by asking children to:
- make up a menu for the café
- choose and price a holiday for two adults, one child aged ten and one child aged two from the holiday brochure
- make a 'till' which will keep different coins and banknotes separately
- make a money box for saving only 10p coins
- make a sorting machine that separates 20p and 50p coins
- find how much it will cost to buy your favourite comic or magazine for a year
- select a new item which might be sold in the school tuck shop and show classmates that it could be profitable for school funds.

What are common difficulties which children encounter and how might these be overcome?

Values of coins – The 2p coin can be difficult for many children as they see one coin with a value of two pennies. More practice with structured number materials, for example, seeing two as one tower of two cubes, one string of two beads, and one piece of orange with two sections.

Place value – Realising the need for the zero in amounts like £2.40, £4.05, £0.67. This might be improved through practice where coins for amounts are set out on a card with columns for £, ten pences and pence.

Notation – Realising that the pound and the pence signs are not used together, for example, 83p or £0.83, *not* £0.83p.

FOUR

Measure

MEASURE

What does measure mean?

Measure involves making comparisons with standards. It is a 'continuous' attribute of an object which is compared, for example, its length or one-dimensional line/path, its area or two-dimensional surface, its volume or three-dimensional space, its weight of molecules. Time is considered as a measure too.

Measurement involves continuous quantities so measures cannot be expressed exactly. Expressions such as 'about', 'nearly', 'to the nearest . . .' are used to illustrate that measures are approximate. The degree of approximation, and so of accuracy, depends on the measuring device and the measurer.

What are real life examples of measure?

Every object, animate or inanimate, natural or man-made has attributes which can be measured.

What is the key vocabulary for measure and what does each word mean?

To compare – to find differences.
Continuous – joined together, not discrete.
Discrete – distinct or separate parts.

Qualitative comparison – relates to the quality being compared, for example, longer, heavier.
Quantitative comparison – comparison made in number terms/units, for example, about 12 rods long, about 62 centimetres long.

What is the historical background of measure?

The Babylonians and the Egyptians developed considerable skills in practical measurement, especially length, area and volume. The Greeks made many of these practical skills a formal study. However, it took many, many years of people using parts of the human body and everyday objects as units of measurement, before systematic use of standard units emerged. Even today, people use different sets of standard units.

What is the value of teaching measure?

Estimation and measurement could be regarded as life skills. It is possible to survive without such skills but little or no knowledge of standard units would make some aspects of communication difficult and involve a dependence on the knowledge of others.

What are possible key steps in development for the learner?

1 Comparisons with 'self' as reference
Initially the children need to develop some measure vocabulary, for example, big, heavy, tall, full, empty. Children could be involved in situations where their own attributes and abilities are used as a reference, for example, when a light switch cannot be reached it is said to be 'high' and when the shopping bag is difficult to lift, it is 'heavy'.

2 Qualitative comparisons between two objects
Adults realise that to say something is 'long' does not have a great deal of meaning. 'Longer than what?' would be the natural response. Children soon discover that they cannot make themselves clearly understood unless a descriptive measure term relates two, or more objects, for example, rather than state that 'Scott's pencil is long', it is clearer to state 'Scott's pencil is longer than Usma's'. A weight might be expressed as 'William has the heaviest schoolbag'.

3 Conservation of measure
Jean Piaget devised the term 'conservation' to help adults realise one of the ways in which children 'see' a situation differently from them. As we mature we reason about what we see, for example, we realise that if we see a car in the distance, it appears about the size of a toy car whereas in reality it is about the same size as one just beside us. This is an example of how we 'conserve' size.

When children are young they believe what their eyes see. In measure, it is important to delay quantitative work until the children can make reasoned comparisons. Reasoning is developed by experience so the children benefit from qualitative comparison activities.

4 Quantitative comparisons
Some form of unit needs to be used if a question such as 'How much heavier is William's bag than Margaret's?' is asked. Measurements can be expressed in arbitrary (non-standard units) and in standard units.

Non-standard units are ordinary objects which are used because they are known to the children and/or readily available, for example, handspans for length, postcards for area and cupfuls for volume. The children can focus on the concept of a unit and be involved in part of the historical development of measure. Activities with non-standard units allow:

- units to be understood as a set of identical objects
- investigation about how units are used for measurement, for example, to find an area, the units have to be laid side by side without leaving spaces or overlapping
- measures to be expressed by counting the total number of units used
- children to find that units don't match what is being measured exactly so results are approximate
- opportunities for measures to be expressed as, for example, 'about . . .', 'just over . . .' and 'nearly . . .'
- estimates to be made as children need to be encouraged to make a reasoned guess based on a previous measure
- different sizes of units to be investigated so that one appropriate to the object can be used
- measurements expressed with whole and fractions of units to give more accurate answers
- children to realise that non-standard units tend to be personal and are not the most suitable for communication.

5 Standard units
Standard units have been created to allow consistency and communication of measures. Some standard units belong to the Imperial system, for example, inches, square yards, cubic feet and miles, and these are not used as widely throughout the world as they once were. Metric units are the system commonly used in Europe

today, for example, metres, litres, grams and square centimetres.

The names of the units can become meaningful to the children as they develop their own references. For example, a metre might be the height from the floor to the pupil's armpit, a square centimetre might be the pupil's middle finger nail, a litre might be the bottle of coke which is bought by the family. Such references are ideal to recall when the child is asked to make an estimate. Estimates should not be wild guesses but reasoned through comparison with a known measure. Children should realise that an estimate is never wrong but with practice they usually make better, closer estimates.

At this stage in the development the pupils learn to use specific measuring devices, for example, rulers, tape-measures, measuring jugs, scales for weighing, and square grids for area. Many of these pieces of equipment need careful explanation and demonstration so that the children understand what is being measured and how the equipment should be used. Often the choice of equipment is based on an initial estimate of the measure. For example, the choice between a 50 cm ruler and a metre stick to measure a length; the choice between a 500 ml measuring jar or a litre one to find a volume; and the choice of weight scales by considering the maximum weight which each can show. The children require practice in using the different equipment. Answers should be expressed in approximate terms such as 'to the nearest . . . (unit name)'.

6 Applications

The children should be challenged to select appropriate equipment for practical tasks, to solve problems and to investigate relationships between units, for example, that one litre of water weighs one kilogram. Such applications consolidate and extend the children's understanding of the concepts of measure and the units involved as well as providing expertise in the use of measuring devices.

What are appropriate resources for teaching measure?

Length
Rods, straws, rulers, metre sticks, tapes, trundle wheels.

Area
Sheets of paper, postcards, envelopes, squared paper, squared acetate grids.

Volume
Spoons, egg cups, cups, mugs, jam jars, coffee jars, litre containers, measuring jugs with millilitre scales, cubes, centimetre cubes and a litre box.

Weight
Marbles, cubes, commercial and home-made weights, groceries with marked gross and/or net weights.

What are possible contexts through which measure might be taught?

Shopping, the supermarket, model making, myself, my room, decorating, and houses.

How might measure be assessed?

Assess *orally* by asking children to:
- name a unit for buying each of the following: potatoes, milk, ribbon, carpet, sweets and petrol.

Assess *practically* by asking children to:
- find which of two boxes has the greater surface area
- price goods for a classroom greengrocer's shop
- find the weight of water in a fish tank
- match a list of measurements with different parts of a tin can (height, circumference, surface area, volume, weight)
- create a data base of measurements of some classmates.

Assess *in written form* by asking children to:

- draw rectangles which have the same area but different perimeters
- draw a bar graph of the results of the rectangles you have drawn.

Assess through *problem solving* by asking children to:

- find the better buy given two or three sizes of the same commodity (chocolate bars, coke).

What are common difficulties which children encounter and how might these be overcome?

Approximate measurements – Many children, and adults, do not realise that quantitative measurement is comparing a continuous attribute of an object with a set of units. Many have difficulty in accepting that all measurement is approximate because ever smaller units could be devised (although these might not be practical). Experience of all types of practical measurement should help such difficulties in understanding.

Fraction – Measurements are often expressed using fractions. The approximate nature of an answer with fractions reinforces that measurements cannot be exact. For example, the length of a pencil could be expressed as 'about 18 cm', or with greater accuracy as 'about 18·2 cm'. If it were practically possible, there could be increasing accuracy to give answers such as 'about 18·23 cm', 'about 18·231 cm', 'about 18·2314 cm' and so on.

LENGTH

What does length mean?

Any line, straight or curved, has a length. Length is a measure of the extent of the line. Lengths can be compared qualitatively to find which is longer and which is shorter. A length can be measured quantitatively which involves comparing it with a number of 'units of length'. Length measurement is often referred to as linear from the Latin word *linea* which means line.

What are real life examples of length?

We use length measurements in our daily lives to: buy clothes; buy carpeting, wallpaper; make furnishings like curtains; and find furniture which is in proportion to a room or which will fit a specific place.

Many people use length measurements in their daily work: tradesmen such as painters, joiners, and plumbers constantly make and use length measurements; surveyors, engineers and architects use measurements to plan and construct buildings and bridges; drivers and pilots need to take account of distances in their calculations of time and speed.

What is the key vocabulary and what does each word mean?

Length is considered to be a dimension, that is, an extent. Breadth, thickness and height are also referred to as dimensions. Each is called linear because it is the property of a line.

There are many descriptive terms associated with linear dimensions, for example:

Big – this is a vague term which does not identify the dimension being referred to, for example, length, area, volume, overall size.

Small or wee – the opposite of big.

Long – is usually used to describe the horizontal dimension of an object. Like the other descriptive terms used for length it does not have a great deal of meaning unless it is used comparatively, that is 'the window ledge is longer than the window', or 'this is the longest of the three ribbons'.

Short – the opposite of long.

Tall – usually used for a vertical dimension from ground level upwards, that is height.

Short/little/small – used as the opposite of tall.

High – usually used to indicate a position above the ground.

Low – the opposite of high.

Deep – seems to be used both for a distance from ground level, or a surface, downwards and to indicate a position below the ground.

Shallow – the opposite of deep.

The conventional units of length which we use are:

Metric

Metre – the basic unit of length, defined at one time as the ten-millionth part of the distance from the North Pole to the Equator along a line of longitude which passed through Paris; now defined by stipulating that the speed of light is 299 792 458 metres per second.

Centimetre – a hundredth part of the metre.

Millimetre – a thousandth part of the metre.

Kilometre – one thousand metres.

Imperial

Yard – the basic unit, historically identified as the length from the tip of the nose to the outstretched middle finger tip.

Foot – a third of a yard.

Inch – a thirty-sixth part of a yard.

Mile – one thousand seven hundred and sixty yards.

What is the historical background of length?

In ancient Egypt man used parts of his body as units of measurement, for example, from the tip of the elbow to the tip of the middle finger was called the cubit and a digit was the width of one finger. Four digits were taken as equal to one palm and seven palms as one cubit.

Units based on parts of the body continued to be used for hundreds of years. Then, in this country, Henry 1 of England decided that such units lacked consistency and decreed that his personal measurement for the yard be used by everyone in the country. In Europe a universal standard was sought. It was decided to divide a quarter segment of an imaginary circle drawn on the surface of the Earth through the North and South poles into ten million equal parts, one of which would be called the metre. The metre is now used widely and has been redefined to achieve greater accuracy.

What is the value of teaching length?

It is an accepted part of communication to be able to:

- describe the dimensions of an object
- make qualitative comparisons
- order by length
- express dimensions in an appropriate unit to a stated degree of accuracy
- estimate dimensions.

People use the units of length to buy material, clothing and other commodities. People use measuring devices like rulers and tapes to take measurements.

What are possible key steps in development for the learner?

1 Language

Young children usually begin by describing the size of objects as 'big' and 'small'. They gradually learn to discriminate in what way an object is big/small and use more specific terms. At first the basic term such as 'long' is used for lengths where a child uses him/herself for comparison, for example, the brush has a long handle if the child has to stretch out his arms to pick it up, the tree is tall if the child has to look up to it, and the light switch is high if the child has to stretch up to reach it.

Although language is the starting place, the development of language of length is continued at each following stage so that the learner's vocabulary is extended.

2 Comparison

Gradually children learn to compare two objects and to state a qualitative relationship between them such as, 'the red pencil is longer than the blue one'. This is usually done by placing the objects to be compared side by side.

3 Conservation

Children are easily deceived by their eyes into thinking that rearrangement can change the length of an object. With a few children it is useful to try out a conservation test:

Show the child a piece of rope with colourful patterns made using a felt pen. Introduce this as 'your snake'. Ask the child to choose from other rope snakes one which is 'as long as' your snake. The snakes should be placed side by side and the child should confirm that one is as long as the other. Tell the child to move his snake 'through the grass' so that the lengths are now out of alignment.

- Ask the child: 'Is your snake longer than mine, is my snake longer than yours, or is your snake as long as mine?' If the child thinks one snake is longer he is likely to be focusing on part of the snakes and making the comparison based on partial information. If the child still realises that both are the same length, he seems to be able to reason that neither length has been added to nor reduced, and has conservation.

4 Ordering

Sets of objects can be ordered according to their length to find the longest/tallest/shortest/widest/narrowest.

5 Arbitrary or non-standard units

Like ancient man, it is useful for children to use personal and other arbitrary units to make measurements. This allows the children to:
- establish that units must all be the same
- find out that units are placed end to end in a straight line to find a distance between two points
- use an appropriate unit, for example a pace, to measure the width of a room and a handspan to measure the edge of a table – this leads to the realisation that a range of units are useful
- express measures to the nearest whole unit or to a specific degree of accuracy, for example, 'almost 5 straw lengths', or 'about $7\frac{1}{2}$ pencils long'
- gain confidence in estimating, realising that a better estimate is usually arrived at by comparing the unknown length with a known length where the same unit was used
- realise that despite the usefulness of arbitrary units, they are limited as a means of communication.

6 Standard units

It is usually metric units which children learn to use in school. The first metric unit can be either the metre or the centimetre. The metre is the more usual choice because:
- it is the basic unit
- it is long enough to allow measurements to be made by laying out metre sticks end to end and to be expressed as a small number
- accuracy is not an important issue.

The centimetre is sometimes chosen because:
- it allows children to measure their possessions such as pencils, books and toys, as well as parts of themselves
- cubes with an edge length of one centimetre can be laid out then counted

- use of the cubes relate well to understanding the centimetre intervals on a ruler
- reading a measurement to the nearest centimetre can be meaningful.

The kilometre is usually taught later. The children could walk this distance and then link it to local distances and map work. The millimetre is also focused on when the children are older, so that drawing and measuring to this degree of accuracy is possible and meaningful. The children are often amazed that aeroplane and house plans are drawn with millimetre measurements specified, until they realise that large and very small items are then measured to the same degree of accuracy. The children meet imperial units such as feet, inches and miles in their environment and the children should be familiar with these and know that they are related to the metric units, for example:

- the yard is just a little shorter than the metre
- five miles is about the same as eight kilometres
- four inches is about ten centimetres.

Children should establish their own references for each of the units, for example:

- a metre could be from the ground to waist/ chest height, or from shoulder to opposite arm finger tip
- a centimetre could be a nail width.

What are appropriate resources for teaching length?

For language and comparison
Sticks, pencils, dolls, cylinders, ribbons, belts

For arbitrary units
A set of sticks of equal length, a length of string with a bead at each end, cubes, a box of new pencils/felt pens/unused crayons, a set of uniform books

For standard units
Metre sticks, tapes marked in metres, centimetre cubes, rulers marked in centimetres, tapes marked in centimetres, trundle wheels, calipers, depth measurer, rulers marked in millimetres.

What are possible contexts through which length might be taught?

There are many possibilities, including: ourselves, toys, planning and building a Wendy house, models, aeroplanes, my room, decorating the classroom, planning and marking out a game/sports pitch/track, sports day, world records, athletics, the Olympics.

How might length be assessed?
Assess *orally* by asking children to:
- estimate distances
- explain how to use a measuring device
- select the best term to complete descriptions of objects involving lengths
- describe a person (footballer/pop star) using as many length words as possible.

Assess *practically* by asking children to:
- measure a table before buying a tablecloth
- measure a bed (in the school medical room) then decide which sheet size is needed
- take measurements of a room for wallpaper and a border
- find out how much curtain material is required for a window
- keep a growth chart of a five- or six-year-old
- keep a growth graph of a tomato plant
- measure a classmate for a knitted jumper
- identify appropriate sizes of clothing for some school children.

Assess *in written form* by asking children to:
- draw a plan to given measurements
- measure features on a plan/map
- carry out length calculations.

Assess through *problem solving* by asking children to:
- design and make a box to keep all their pens, pencils and erasers in
- investigate the number of stitches which make 2 centimetres for different needle sizes and wools

- investigate the relationship between shoe sizes and length of feet.

What are common difficulties which children encounter and how might these be overcome?

Estimates – Children often lack the confidence to guess because they don't want to be wrong. They should be encouraged to relate unknown distances to known ones, for example, the window ledge is just a little longer than the window; the book length is about half the desk length. Estimates should never be regarded as 'wrong', some are simply better than others. The challenge can be set for children to become better at making estimates through practice.

The approximate nature of measure – It is difficult to understand that measurement cannot be exact. Children should be encouraged to state all measure answers as 'about' or to the nearest unit.

Units – Through practical experience the children should establish references for the different units and be able to relate units to each other. This requires lots of practical tasks, both in a group where there is the opportunity to learn from others and individually.

Equipment – Children often use equipment incorrectly, for example, using a ruler by beginning at the mark labelled 1; not knowing how to use a 150 cm tape to measure the length of a table about $2\frac{1}{4}$ m long; not knowing how to set a trundle wheel so that the first 'click' marks a full metre length. A new piece of equipment may be demonstrated or, if the children investigate how to use it themselves, the effectiveness of their use assessed.

AREA

What does area mean?

Area is the measure of the extent of a surface. To find the measure of an area involves reckoning the number of unit squares within its boundaries.

What are real life examples of area?

People are interested in the floor area of their home, the area of each room, the area of the walls and of the garden. Area is the concern of the farmer as by working out the area of his land he can work out how to use his land in the most efficient way. Area is the concern of the map and the plan makers. Area is an accepted form of communication about the environment.

What is the key vocabulary for area and what does each word mean?

Some abstract nouns include:

Surface – the exterior part of anything, any continuous two-dimensional shape.

Surface area – the complete extent of the exterior part of a three-dimensional shape.

Descriptive terms tend to be:
Greater, larger, smaller.

Units of area are:

Metric
Square metre, square centimetre, square millimetre, square kilometre, hectare.

Imperial
Square yard, square foot, square inch, square mile, acre.

Relationships between the metric units are:

$$100 \text{ mm}^2 = 1 \text{ cm}^2$$
$$10\,000 \text{ cm}^2 = 1 \text{ m}^2$$
$$1\,000\,000 \text{ m}^2 = 1 \text{ km}^2$$
$$1\,000 \text{ m}^2 = 1 \text{ hectare}$$

What is the historical background of area?

Egyptian surveyors measured area. The Babylonians and the Chinese also used area. It is not known who was the first to consider area in terms of square units, but it may have arisen from a mosaic floor or even from basket work. The Greek mathematician Antiphon attempted to find the area of a circle by drawing a square inside it, then by drawing a regular octagon, then a regular sixteen-sided shape and so on until he had drawn a shape which approximated the circle itself. Archimedes improved on this method. He also found a method of finding the area of an ellipse.

The imperial unit of an acre was established as the amount of land which a pair of oxen could plough in a day.

What is the value of teaching area?

Area is used to convey the size of surfaces so it is useful to have a concept of each of the main units.

What are possible key steps in development for the learner?

1 The concept of surface

Young children in nursery and infant classes are involved in painting, in potato printing, in ironing, and in covering a bed or a table with a cover. All these experiences where they are covering surfaces, provide opportunities to establish an initial concept of area.

2 Direct comparison

Making comparisons of the surfaces of objects by placing one on top of the other is a useful experience which can lead to phrases like 'is larger', 'has a greater surface', 'covers more table', 'has more surface', and 'has a smaller area'.

3 Conservation of area

A few children aged six or seven can be tested to find if they have grasped conservation of area.

Ask each, individually, to identify from a choice of three, a sheet of paper which covers the same amount of table as a sheet which you call 'my sheet'. The child then cuts up 'his' or 'her' sheet into two pieces. Ask if your sheet covers more table, if his/her sheet covers more table, or if both sheets cover the same amount of table. The child is required to reason that the sheet which has been cut into two pieces has not had its area changed, so it still covers the same amount of table.

4 Arbitrary or non-standard units

If a question such as 'How much greater is the surface of the workbook than the storybook?' is asked, arbitrary units could be used to find a quantitative answer. The children should realise that units of area must themselves cover a surface, for example, postage stamps, postcards, triangles, counters, playing cards and mats. The area is found by the children counting the total number of units within the shape outline. A decision has to be made frequently whether to count a part unit, and a rule such as 'count only the part units which are half or more than half' might be devised by the children themselves.

Arbitrary units are useful to establish that:

- an appropriate unit is used for each size of surface to be measured
- measuring with a unit involves using a number of the same item, for example, if an envelope is the choice of unit, a number are required which are the same shape and size
- the units are placed side by side without overlapping over the surface to make the measurement

- the result of the measurement is approximate.

However, arbitrary units are limited in their communicative value and standards understood by everyone are required.

5 Standard units

The square centimetre is usually the first metric standard to be used by the children. Centimetre cubes or an acetate squared grid can be placed over a given surface, or the object can be laid on top of centimetre squared paper and an outline drawn, so that the area can be counted as a number of square centimetres. Answers will be expressed to the nearest square centimetre.

The children can also be given the task of using centimetre squared paper, or a nailboard with a 2 cm square grid, to create different shapes with the same area. References for a square centimetre of area could be a finger nail or a blouse or shirt button.

A square metre of area can be made with newspaper or wrapping paper. Although the easiest shape to make is the square, more than one should be made so that some can be cut into two or three parts which are rearranged and taped together to give a range of different shapes. The important learning point here is that one square metre of area can take many shapes, whereas a one metre square must be a square with an area of one square metre. A reference for children for this area might be the surface of the teacher's table.

The square kilometre can be illustrated on a local map so that the children can establish a real reference for a unit of this size. The square yard could be illustrated in a situation where this can still be used, for example, in carpeting. The square yard can be compared with the square metre and found to be a smaller area.

6 Formulae

Ideally children should discover a formula for themselves. It should be regarded as a short method of finding an answer. In area work the children may realise as they count squares to find the area of a rectangle that it would be quicker to find the number of squares in one row and multiply this by the number of rows. In the same way, the children might find a formula for calculating the area of a right-angled triangle as this shape can be seen by them as half of a rectangle. Wherever possible, examples should be related to areas of places, playing fields, ponds, gardens, flower beds, tables, book covers, sheets of different papers, and jewellery to give a meaningful context for the calculations.

What are appropriate resources for teaching area?

Stamps, postcards, flat shapes, envelopes, and sheets of paper make suitable non-standard units. Squared paper, with squares of edge 1 cm, 2 cm and 5 cm, is useful. Squares of 10 cm by 10 cm, possibly the 'flats' from the structured number representation set are useful when finding larger areas. Rulers, metre sticks, surveyor's tapes, paper, and maps may also be used.

What are possible contexts through which area might be taught?

The garden, the farm, building a house, the swimming pool, the football pitch, decorating, and making jewellery.

How might area be assessed?

Assess *orally* by asking children to:
- play a game where each person names an object with a slightly larger area, beginning with a pinhead
- estimate the area of the door and the floor.

Assess *practically* by asking children to:
- find the area of the car park
- find the area of the assembly/dining hall
- find the approximate area of a pair of shoes.

Assess *in written form* by asking children to:
- calculate the area of plots drawn on a map

- calculate the area of each room in a house plan.

Assess through *problem solving* by asking children to:

- find a person's approximate body surface area
- design a class badge and then write an order for the area of metal required to make one for each person in the class
- find the area of the school building and the school site.

What are common difficulties which children encounter and how might these be overcome?

Perimeter/area – Confusion can arise between perimeter and area. However, if the children have had enough practical experience to find that length is a one-dimensional attribute and area a two-dimensional one, differences should be obvious. They should also be aware that each is expressed in different standard units.

WEIGHT

What does weight mean?

The amount of matter in an object is called its mass. Weight is a measure of the force which the Earth exerts on a body. The weight of a body will change depending on how far it is from the Earth's surface. Astronauts coping with weightlessness in space illustrate that their mass does not change while their weight does. Newton's second law relates weight and mass as: weight = mass × gravity (g = 9.81 ms^{-2}).

Weight should be measured in newtons, but convention has established that we use the units of mass, that is, kilograms and grams.

What are real life examples of weight?

By law, the weight of the contents of packets, bags and tins of food and other commodities is recorded for the shopper to read. Recipes involve the cook in weighing out ingredients. Lifts, bridges and floors are designed and constructed to carry stated weights. Transport vehicles have an unladen and a maximum loaded weight. Air passengers are limited to a specified weight of baggage, so that the overall weight of the aeroplane allows it to get airborne safely. Babies have their weight recorded at birth and most people keep a check on their

body weight throughout their lives. A sudden loss or gain in weight may indicate an illness. Postage costs are determined by weight.

What is the key vocabulary for weight and what does each word mean?

Abstract nouns include:

Weight/mass – the association of a number with a unit of weight to describe the amount of matter which makes up an object.

Some weighing equipment includes:

A balance – a machine which indicates whether two objects: weigh about the same or balance; have one object heavier; or have one lighter than the other.

Scales – a weighing machine where units of weight are represented by regularly spaced marks and the weight of an object is indicated by the position of a pointer on these marks.

Some weighing equipment is much easier to use as it gives a digital display of the weight. Such machines are gradually being used in classrooms.

An action verb includes:

To weigh – to determine how heavy an object is.

Two descriptions of weight are:
Heavy and light – qualitative terms used to compare the amount of matter. They require a reference for comparison to be meaningful, although young children tend to consider an object heavy if they find it difficult to pick up.

Other terms include: **heavier**, **lighter**, **heaviest**, **lightest**.

Units of weight:

Metric
Kilogram (kg) – the standard unit of mass – It is defined as the weight of 1 litre of water.
Gram (g) – a unit of mass which is the weight of one millilitre of water. There are 1000 g in 1 kg.

Imperial
Pound (lb) – 2.2 pounds are about 1 kilogram
Stone – there are 14 pounds in 1 stone.

What is the historical background of weight?

The balance was referred to as early as 3500 BC. People wanted to ensure fair trading but they did not find this easy as they could not decide on a standard measure. However, the Egyptians used a thousand grains of wheat as their basic measure. They made objects, often in the form of small sculptures that were as heavy as the thousand grains and used these as weights. Weights were made for fractional parts of the basic unit such as 200 grains.

In 1351 AD, Edward III devised the standard weight, the stone. The original stone can be seen in Winchester City Museum. Stones and pounds are imperial units of weight and are still used. More universal are the metric measures which were invented by the French in the eighteenth century. The piece of metal which was chosen as the kilogram can still be seen in Paris. The kilogram is equivalent to 15664 grains of wheat and so is about $15\frac{1}{2}$ times the basic unit used in Ancient Egypt.

What is the value of teaching weight?

It is useful to have a concept of weight and of the main units used as well as to be able to:
- recognise and request weights of food
- interpret the scale on a weighing machine
- know the relationships among the units of weight.

What are possible key steps in development for the learner?

1 Language and comparison
Weight needs to be brought to many children's attention as it is not an attribute of an object which they can see. They should learn to pick up objects, to feel their heaviness. At first the descriptive terms that young children use are 'heavy' and 'not heavy'. Gradually they learn that comparison of two objects leads to more meaningful terms like 'heavier' and 'lighter'. At this stage in the development the balance is introduced so that children are not making the comparison of weight by guessing. They could be shown that the bar of the balance is a see-saw where, if it is horizontal the two objects are about the same weight, or if one end is lowered that this indicates it is supporting a heavier object. They should learn to observe the bar at eye level. Children could order three objects by weight. This cannot be done directly if the balance is used, so the children need to establish relationships between pairs and then reason the order.

2 Conservation
Young children are influenced by what they see and the teacher should find out if this applies to their understanding of weight. Two pieces of Plasticine could be compared and found to be about the same weight. The children should then change the shape of one of the pieces. Some children will believe that the more spread out shape now has a greater weight. Children require the maturity to reason that nothing has

been added or taken away from each piece and so both are still about the same weight. If the child cannot do this, it would seem that he or she cannot conserve.

Many children require a great deal of experience to dissociate weight from size and to accept that large objects can be light and small objects heavy.

3 Arbitrary units

Children should choose and use suitable everyday objects as units to quantify weight. Results should be expressed to the nearest unit which achieves a balance, for example, 'about 6 nails', 'nearly 10 marbles' so that they are becoming aware of the approximate nature of weight. They might also discover that the same object is balanced by fewer heavier units than lighter units. Once they have had experience of a particular unit, the children can estimate the weight of an object. Estimates should be reasoned as heavier or lighter than a known object.

4 Metric units

Children should realise the need for units of weight which can be universally communicated. The kilogram is the basic metric unit used for weight. It is sensible to begin with this as children find it difficult to 'feel' the lighter gram unit. There are not many objects in the classroom which weigh about or more than 1 kilogram and which can be placed on the pan of a balance, so the teacher could prepare some bags of stones, bundles of monthly magazines, some reference books and some boxes of nuts and bolts or nails. Ideally the children could make their own kilogram bags of sand or pebbles and these can then be divided to give a half and two quarter kilogram weights. Objects can now be classified as less than $\frac{1}{4}$ kg, more than $\frac{1}{4}$ kg but less than $\frac{1}{2}$ kg, more than $\frac{1}{2}$ kg but less than $\frac{3}{4}$ kg and so on. Children could focus on 100 g as a familiar weight because many items are packaged in 100 g (and multiples of this)

and there are suitable resources of about 100 g in the classroom for them to handle.

5 Interpreting scales

Many weighing devices have scales which the children need to interpret. Bathroom scales usually show a simple scale with the main intervals being in kilograms. As scales vary so much children should note:
- the heaviest weight which can be recorded (and to weigh objects within that limit on these scales)
- the weight represented by the main interval
- the weight represented by each smaller interval
- the accuracy with which a weight can be expressed (usually to the nearest small interval mark, but sometimes halfway between these marks can be meaningful).

6 Net weight and gross weight

Children should be aware of the difference between the overall or gross weight, and the net weight of the contents. Grocery packaging should be collected to investigate these weights and the differences involved.

What are appropriate resources for teaching weight?

A two-pan balance, a variety of arbitrary and metric weights, a variety of scales (bathroom, kitchen, letter, angler's etc.), packets to weigh, grocery items to read their weights, recipes.

What are possible contexts through which weight might be taught?

Projects might be a classroom shop – butcher, greengrocer, and a supermarket where some items are priced by weight. The post office, the airport, the restaurant, the bakery, the hospital, and myself are also useful contexts. The children could bake or make up recipes which do not require cooking. A visit could be made to a weigh bridge.

How might weight be assessed?

Assess *orally* by asking children to:
- explain how the two-pan balance works
- describe the steps they take to 'set up' a balance or scales and use them
- estimate the weight of an object in arbitrary/metric units.

Assess *practically* by asking children to:
- identify the heavier/lighter/heaviest/lightest object
- find the weight of an object in arbitrary/metric units using a two-pan balance
- find the weight of an object using scales.

Assess *through written examples* where:
- net/gross weight is calculated
- weights are expressed in different forms, for example, 3465 g is also 3 kg 465 g and 3.465 kg.

Assess through *problem solving* such as:
- finding the best buy by comparing different weight packages of the same commodity
- creating a recipe

- constructing a two-pan balance
- constructing a spring balance
- finding the average weight of a ten-year-old boy/girl
- making a 50 g Plasticine weight (using only a 20 g weight and a balance)
- dividing Plasticine into three equal pieces using a balance.

What are common difficulties which children encounter and how might these be overcome?

Feeling weight – Children find it difficult to estimate and compare weights when holding objects in the palm of their hands due to the area of contact varying. It is easier when an object is placed in a plastic bag and suspended from the fingertips.

Balances and scales – Children should be taught to make a balance level and to 'zero' scales before using them. They should also be aware that many weighing devices become less accurate or faulty through moving them about or misusing them. Sometimes a lump of plasticine is stuck to a pan to make it balance.

VOLUME

What does volume mean?

Volume is the measure of space taken up by a three-dimensional object. The space within a container is known as its capacity but as the thickness of many containers is negligible, it has become acceptable to refer to the space inside as volume too.

What are real life examples of volume?

Liquids are often bought by volume, for

example, a litre of lemonade, 50 litres of petrol, 2 litres of paint, a pint of milk. Ice-cream can also be bought by the litre.

The number of fish which can be kept in a tank, either at home or at the fish farm, is calculated by knowing the volume of the tank and the required living space for each fish. In the same way, people are allocated working space when buildings are planned and built. A comparison of old and modern houses shows how the amount of space allocated for different functions has changed for example, modern houses have more but smaller rooms.

What is the key vocabulary for volume and what does each word mean?

Some abstract nouns include:
Space – an amount of surface with height, it has three dimensions.
Displacement – the amount of liquid which is moved when a solid is submerged; this amount can be collected as overflow (a special displacement container can be used) and will be found to equal the volume of the solid.

Some volume equipment includes:
Measuring jug – can be of different volumes with a variety of millilitre scales allowing volume to be measured to varying degrees of accuracy.
Centimetre cubes – can be used to pack boxes to find how much space is inside.

Arbitrary units of volume could include:
spoonfuls, **cupfuls**, and **marbles**.
Conventional units include:

Metric
Litre is the basic unit.
Millilitre, 1000 ml = 1 litre
Cubic centimetre, $1 \text{ cm}^3 = 1$ ml

Imperial
Gallon, 1 gallon is about 5 litres
Pint, (8 pints = 1 gallon so the pint is about $\frac{3}{4}$ of a litre).

Some descriptive terms used for volume include:
holds more   holds less
holds about the same amount
greater   smaller
full   empty

Some formulae for volume include:
Volume of a cuboid is
length × breadth × height.
Volume of all prisms is
area of cross-section × height.
Volume of all pyramids is
one-third the area of the base × height.
Volume of a cube is
length^3.

What is the historical background of volume?

A papyrus, now in Moscow, suggests that the Egyptians possessed the correct formula for the volume of a pyramid about 1800 BC. The Greek mathematician Democritus gave the correct value for the volume of a pyramid too, and explained his method as seeing it made up from a large number of slices. Another Greek, Archimedes, realised that a floating body displaces its weight by an equal amount of water.

What is the value of teaching volume?

A more comprehensive understanding of our environment and the objects in it can be made with a knowledge of volume. Here are three simple examples:
- it is sensible to consider not only the length, breadth and height when buying a new refrigerator but also its internal capacity
- a glass of lemonade should not be filled too near the top if ice is to be added to it
- the volumes in two containers of liquid detergent could be checked as the shape of a container can be misleading, regarding the volume of the contents.

What are possible key steps in development for the learner?

1 The concept and language of volume
As with other measures, children require practical experience to begin to form the concept of an object taking up space. As children fill boxes with bricks and pour liquids into containers, they develop language such as 'It can hold more', 'It's full', and 'There's not any space left'.

2 Comparison and ordering
Containers can sometimes be compared perceptually. Some comparisons can be made by placing one container inside the other. For most

comparisons, however, pouring from one filled container to the other is required.

3 Conservation

A test can be carried out to see if children realise that two matched amounts of liquid remain the same when one amount is poured into a container of a different shape, or two matched shapes made with linking cubes have the same volume when one of them has the cubes rearranged so that the shape is altered.

4 Arbitrary or non-standard units

When the comparison is made to find out, not only which container holds more, but also how much more, a unit of volume requires to be used. Children gain a better understanding of what a unit of volume means, if they begin by using arbitrary units like eggcupfuls and tubfuls. These names probably should be of the form 'tubsful' but convention has placed the 's' at the end. When using arbitrary units, children should learn to express volumes as 'about' to emphasise the approximate nature of measure.

5 Standard units

These units could be made meaningful by looking at the volumes of commodities. For example, the bottle of lemonade for the litre, the milk carton for the half litre, the container of fresh orange for the quarter litre and the tub of yoghurt for 100 millilitres. A cubic metre could be built and compared with spaces such as that under the teacher's table or in a cupboard. Imperial units should also be considered, for example, the pint of cider or beer and the gallon of garden fertilizer.

The children should be able to 'read' volume scales to be able to measure quantities of liquid. They need to be able to say what the main and the smaller intervals on a scale represent, and realise the maximum volume and the accuracy of the readings of different measuring jugs. Volumes should be expressed in the form '320 ml to the nearest 20 ml'.

6 Relationships among standard units

Children could find out that a plastic box which holds 1000 centimetre cubes can also hold 1 litre of water and so deduce that one cubic centimetre and 1 millilitre represent the same amount of space. It is interesting for them to see one millilitre of water spread over a surface and compare this with a centimetre cube. It is difficult to believe that they are the same volume. An investigation could be carried out to find that 1 litre of water weighs 1 kilogram.

What are appropriate resources for teaching volume?

Bottles, cartons, containers, measuring jugs, funnels, liquids, dry sand, small boxes, centimetre cubes, a plastic litre box.

What are possible contexts through which volume might be taught?

Projects could include: making an animal house or an aquarium, building a house, making a garden pool, the café and ourselves.

How might volume be assessed?

Assess *orally* by asking children to:
- explain if they have a volume
- explain how to take a reading on a given measuring jug.

Assess *practically* by asking children to:
- put containers in order of volume
- make a card container which will have the same volume, and so just hold a given pile of sand
- build three different shapes with the same volume using interlocking centimetre cubes
- follow a dessert recipe which involves liquids, for example, making a jelly.

Assess through *written examples* such as:
- A fish tank has a length of 800 mm, a breadth of 300 mm, and a height of 400 mm. It is to be filled three-quarters full of water. What

volume of water will be used and what weight is this?

Assess through *problem solving* by:

- creating a recipe for a fruit drink
- finding the best buy among three sizes of coke
- putting containers into sets according to their volume.

What are common difficulties which children encounter and how might these be overcome?

Realising which of two containers holds more – If children pour from a larger container to a smaller one when comparing them, they will find they have a full container and another with only a small amount of liquid in it. It can seem, in answer to the question 'Which container holds more?' that the full container is holding more at that moment. Some young children find it difficult to remember that the container they began with held all the water. Encourage children to estimate which is the smaller and

pour from it to the larger when making a comparison. They could talk through what they are doing and what this means.

Estimating volume – It is difficult for anyone to estimate volume. If you ask children to do this, encourage them to compare the unknown container with one they know, but also to allow about 20% as an error factor.

Reading a volume scale – It has been known for children to hold a ruler beside a cylindrical container and to try to read the volume of its contents. They have not grasped that a measuring jug scale is measuring three dimensions, not just height. It might help if they made their own measuring jug using a 100 ml container.

Formulae – should be delayed until the children have had sufficient practical experience to understand how the dimensions must be expressed in the same units and how the units change as they are multiplied or divided, for example: A box is $\frac{1}{4}$ m long, 20 cm wide and 50 mm high. The volume of the box is 25 cm $\times$ 20 cm $\times$ 5 cm = 2500 cm^3.

TIME

What does time mean?

Time is a measure. It is difficult, however, to express in words what is being measured – is it 'existence' or 'from one moment to the next'? Time cannot be seen or touched but we are surrounded by the effects of time passing, as each night follows day, and one season leads to another.

There are two main ways of considering time:

- a 'moment' which is recorded, for example, as 3:05 and as 'five minutes past three'
- a 'duration', for example, the afternoon, a day, a century, an hour and a half.

Calculations of time are durations.

What are real life examples of time?

Many people believe their lives are controlled by time – time to get up, time to be at work, time to eat, time to catch the train and so on. Certainly no one can be unaware of time with clocks showing the time of day and the calendars showing the day, month and year, as constant reminders to us that time is passing.

What is the key vocabulary for time and what does each word mean?

Some abstract nouns include:

Duration – the amount of time something lasts.

Moment – a very short duration of time.

Instant – so short a duration that the beginning and the end are almost the same.

Some time equipment includes:
Sandtimer – an early way of measuring time where a specific amount of sand takes a stated time, like one, three, five or sixty minutes, to run from one glass or plastic chamber to another through a narrow neck.
Analogue clock/watch – time is displayed using a continuous scale, usually circular, with two pointers or 'hands' of different lengths; the scale usually shows 12 hours and the shorter hand travels around this twice each day, while the longer hand indicates the minutes on the same scale and makes a complete rotation every hour, completing 24 rotations each day.
Digital display – time is recorded by a display of numerals which represent the hour and the number of minutes past the hour; the display can be based on 12 hours or 24 hours.

Time is expressed in imperial not metric units. Some units of time include:
Day – the time the Earth takes to make one revolution on its axis, from midnight to midnight; any consecutive 24 hours can also be labelled as a day.
Hour – the day is divided into 24 hours.
Minute – each hour is divided into 60 minutes.
Second – each minute is divided into 60 seconds.
Week – any seven consecutive days.
Month – from one new moon to the next, usually 30 or 31 days but may be 28 or 29 days; any consecutive four weeks can be referred to as a month.
Year – the time taken by the Earth to move around the sun, accepted as 365 days.
Leap year – the actual time taken for the Earth's rotation is $365\frac{1}{4}$ days so every fourth year (those exactly divisible by 4, for example, 1992 and 1996, except century years which must divide exactly by 400), there are 366 days (the extra day is designated 29th February).

Season – the year is divided into four seasons.

Some descriptions of time include:
Long/short – the apparent length of time is influenced by how it is spent.

An action verb is:
To time an event – to measure how long it takes.

What is the historical background of time?

Time and travel were often linked in the past. Distances were expressed as 'two moons away' or 'from sunrise to sunset'. The first measurement of time was based on natural events which recur at regular intervals. In Babylonian times the people created a 'calendar' of 360 days by considering the pattern of night and day and the seasons. The Egyptians believed this required to be adapted and added five feast days. Later a Greek astrologer, Sosigenes, conceived the concept of the leap year to make the year a more accurate measure. This was used in the Julian calendar which began in 46 BC. The calendar was reformed in 1582 by Pope Gregory XIII and his Gregorian calendar was adopted in Britain in 1752 and is still used today.

It is interesting to note that the end of a century is not, for example, at the end of 1900, but at the end of 1901 because there is no year labelled 0. The year 1 AD follows the year 1 BC directly. Short durations of time have been measured by interpreting the direction and length of a shadow cast by the sun, by collecting drips of water, by letting sand run from one container to another, by swinging a pendulum, by rocking a cog wheel and by burning a candle.

What is the value of teaching time?

To be able to communicate and interpret times and dates are important life skills.

What are possible key steps in development for the learner?

1 Early vocabulary and sequencing events

Unlike other measures, there is pressure to rush towards standard units, for example, for reading the time of day, to be able to name the days of the week and the months of the year, before the concept of time is grasped. This is an emphasis on the 'moments' of time. However, it is important to help the children to understand the concepts associated with the duration of time and to involve them in experiences about sequencing events so that they can consider the words 'before' and 'after'. They could consider initially what comes 'after', using events like going to bed, putting on your socks, opening a door. Then the more difficult concept of 'before' could be tackled in the same way. A child could be given a short routine to perform and the other members of the group asked 'What did she/he do before she/he did . . .' and 'What happened after he/she . . .' Daily happenings, which should be understood as intervals of time, can be ordered. Other language at this stage could include 'now', 'soon', 'bedtime', 'storytime', 'playtime', and 'hometime'.

2 Reading the time

The digital display is easy to interpret. It shows both hours and minutes together. The children can regard it as an instrument which counts the minutes and the hours.

The analogue clock face is much more difficult to interpret and understand. The children are reading an hour time line which is usually circular with the numeral 12 placed so that the line seems to have no beginning or end. The children can be introduced to the hour hand by using it alone to read 'three on the clock' or 'three o' clock'. The children can state, for example, if the hour hand is between 3 and 4, that it is after three o'clock but before four o'clock. Later the minute hand can be introduced to read more accurate time, combined with the hour hand. An analogue clock face with one hand and a visual aid of digital hour times could be used to introduce the o'clock times. It can be explained that it is only three o'clock for an instant but there is a whole hour until there is another instant called four o'clock. The children can see on an analogue clock that the hour hand moves very slowly. On the digital display, the 00 alongside the hour number changes as 'an hour is counted'. It is difficult to give a child a concept of one hour but references can be established for events that last for an hour by setting a timer to ring after an interval of an hour.

It is much easier to give children the concept of one minute by setting them tasks like 'How many times can you button and unbutton this cardigan before the sand in the minute timer runs out?' The minute hand can be linked to the movement of the hour hand on the analogue clock and the numbers which count the hour on the digital clock can now be explained as minutes.

The numbers beside those of the hour on the digital display are explained as minutes past the hour, for example, 10:35 is described as thirty-five minutes past ten. This terminology is recommended for the analogue clock too. The phrases 'half past' and 'quarter past', usually associated with the analogue clock can be accepted as another way of saying 'fifteen minutes past' and 'thirty minutes past'. The phrases 'minutes to' and 'quarter to' the hour are best left as late as possible so that such a convention can be understood and the difficulty of using the 'following hour' coped with, for example, 6.40 is read as 'twenty to seven'.

3 Days and months

Children learn the sequence of the days of the week but sometimes don't fully understand the way in which the names are used over and over again. They also learn that any seven days can be interpreted as one week and not only Sunday to Saturday. The terms 'today', 'tomorrow', 'yesterday' and 'weekend' could be taught at this stage. A weekend is usually accepted as Saturday and Sunday, although a 'long' weekend might include Friday or Monday as well.

The sequence of months can also be learned. The grouping of the months into four seasons could be related to work on the changing weather and environmental conditions.

4 Timing events

It is important that the concept of time as a duration is emphasised from the beginning. For example, if Andrew is having lunch he does not have it at one o'clock but during a time after one o'clock until about half past one. Children could time class and leisure activities in minutes using a stop-watch, preferably with a digital display.

Children should be encouraged to estimate how long they think an event or activity will take, and how long they spent at something. They should also realise that time seems short if they are interested and busy, and long if they are bored or just listening.

5 Twelve hour notation

The terms 'morning', 'noon' and 'afternoon' and the abbreviations am and pm should be related to the 12 hour notation.

6 Twenty-four hour notation

This work could be introduced through a context which is particularly meaningful, for example, setting an alarm or a video recorder, timetables for travel and holidays. A time-line from midnight to midnight where 12 hour and 24 hour notations are both recorded is a particularly useful learning device.

7 Calculations

Calculations where the starting time, the finishing time or the duration is to be found could be carried out using a time-line. The method of 'counting on' is recommended,

for example:

The TV programme begins at 3:45 and ends at quarter past five. How long does it last?

8 The second

Children could learn that each second can be counted by saying a phrase like 'one elephant, two elephants, three elephants . . .' as well as by observing the seconds hand on some analogue clock faces or the digital seconds display of some watches. Activities could be timed and the average time found, for example, for writing your name. Sports events and games on television often show a digital display of seconds and sometimes tenths of a second so it would be interesting to discuss these.

9 Time-lines and time zones

Work on time zones could be developed from contexts such as travel and holidays. The concept of time zones could be made meaningful by a pupil, or the teacher, telephoning a relative where it is discovered that he/she is just getting up while for the pupils here in the United Kingdom, it is afternoon.

Year time-lines are excellent to help children achieve a sense of sequence for the past. Historical projects benefit greatly from the children making an appropriate time-line. In this way, too, 'decades', 'centuries' and 'generations' can become meaningful terms.

What are appropriate resources for teaching time?

Appropriate resources include: watches, clocks; newspapers, magazines; postmarks, calendars, diaries; timetables for school, events, bus, train, plane and ferry; tickets, adverts, posters and programmes; holiday brochures.

What are possible contexts through which time might be taught?

The airport, the journey, holidays, the post, newspapers, car racing, athletics, skiing, swimming, and the Olympics.

How might time be assessed?

Assess *orally* by asking children to tell you:

- the time on an analogue clock
- the time on a digital display
- the day and date
- familiar times in their daily/weekly routines
- days and times of well-known TV programmes
- the time of a programme from a daily TV guide
- the day and time of a TV programme from a weekly guide
- the time of a plane/train/bus/ferry from a timetable
- the return date of a library book after a two week loan
- a duration/starting time/finishing time which can be calculated mentally.

Assess *practically* by asking children to:

- set the hands of an analogue clock
- show a digital time on a visual aid
- set an alarm clock
- set a time on a video recorder
- interpret a postmark
- find the date of a missing magazine in a weekly or monthly sequence
- time an event
- use this year's calendar to find
  - the date of the first Tuesday in June
  - the day and date of one week after May 2nd
  - the day Christmas will be.

Assess through *problem solving* by:

- making a digital display
- creating a device to time one minute
- designing a postmark
- finding how much Mrs Dunn's papers will be

for the month of March this year if she gets *The Guardian* each day and *The Observer* on Sundays

- finding the cost and the time taken for different ways that you could travel to London.

No examples for written assessment have been included as the use of resources make the emphasis in most examples practical or oral. Examples in a textbook could be used as written assessment.

What are common difficulties which children encounter and how might these be overcome?

The hour hand on an analogue clock – Some children do not realise that this moves and that you can tell the approximate time by it alone. Let them practise reading times using a face with only the hour hand.

When two hands are used some children need specific practice in realising which hour they relate 'half past', 'quarter past' and especially 'quarter to' with. Again practise reading such times using only the hour hand.

Relating digital and analogue times – It might help if children could have a visual aid of a number strip showing 01, 02, 03 . . . 58, 59, 00. The strip could be halved to show that 30 is half way, 15 is at the end of a quarter of the strip and 45 at the end of three-quarters.

The continuity of time – Some children find it difficult to state what is on TV at a specific time, as only starting times for programmes are listed in TV guides. It is important that children consider the durations of programmes and can identify what would be on the TV channels at any moment from consulting a daily guide.

A week later – Children should realise that the seven days are counted from the next day. They could also be shown how you can move along a row or down a column to find a week later on a calendar. A week earlier should be considered too.

The number of minutes in an hour – Some children think there are 100 minutes in an hour instead of 60 minutes. This is probably because so many measures are metric and the expression 'fourteen hundred hours' is sometimes used. Children should be clear that time units were devised long before metric units and that they have unusual relationships. Explore these relationships:

> 60 seconds = 1 minute
> 60 minutes = 1 hour
> 24 hours = 1 day
> 7 days   = 1 week

A time such as 14:00 should be read as 'fourteen zero zero hours' or 'fourteen oh oh hours'.

Counting on – When finding durations, for example, 'When does a TV programme end which begins at 8.25 pm and lasts for 1 hour 40 minutes?' Some children will find it easier to begin with the starting time, add any hours and then add the minutes. For instance; 8.25 to 9.25 is 1 hour, 9.25 to 10.05 is 40 minutes, so the programme finishes at 10.05.

SCALE

What does scale mean?

When a measurement is represented either larger or smaller than its true measurement, it is said to be scaled. The relationship between the represented measurement and the true measurement is expressed as a 'scale factor'. If the scale for a drawing of a square was 1:2, each length measurement shown is one half of the real measurement; or 1 cm on the drawing represents 2 cm on the real square.

What are real life examples of scale?

Plans and maps are scaled representations of true areas. Different scale factors can be clearly seen if you look at a road as it is represented on a map of the country, a map of the region, and a map of the town.

    Scale models of vehicles, animals, buildings and people exist as toys and as souvenirs. Architects and engineers make scale drawings, and sometimes scale models, of buildings and structures they plan to build. Photographic processors make reductions and enlargements. Telescopes and binoculars are used to give enlargements.

What is the key vocabulary for scale and what does each word mean?

Reduce – make smaller.
Reduction – the result of making something smaller.
Enlarge – make larger.
Enlargement – the result of making something larger.
Scale – the amount of reduction or enlargement, for example, the lengths of an item can be doubled, multiplied by 4, by 10, by 50, halved, divided by 5, 10, 32, 64, 100.
Scale factor – the relationship between the true length and the represented length can be expressed as a ratio. For example, toy model cars are often 1:32 or 1:64. This means that each centimetre length of the model represents 32 centimetres (or 64 centimetres) in the real

car. 1:48, 1:72 and 1:144 are usual scale factors for model aeroplanes.

What is the historical background of scale?

Ancient mathematicians knew that if a stick is placed vertically when the sun is 45° above the horizon, the stick's shadow is of equal length to the stick itself. This principle was used to find the heights of objects like trees and buildings. However waiting for the sun to be at a 45° angle was not practical so this idea was developed to relate the length of the shadow to the length of the stick at any one time and use this 'scale factor' to find unknown heights from the length of their shadows.

What is the value of teaching scale?

Scaled drawings are the accepted form of communication for plans, maps and production sketches, so it is useful for children to be able to interpret the true measurements from a map and to create scaled representations as they draw plans.

What are possible steps in development for the learner?

1 Representations

Young children learn to accept representations of real things – they talk on their toy telephone instead of making real calls, they point to a picture in a book and name it as if it was the real thing, and they play with a toy tractor in the same way that the farmer might drive the real tractor on the farm. A good beginning for work on scale could be to discuss the ways in which pictures and toys are like and unlike the real objects which they represent.

Children could choose boxes to represent buildings, for example, a row of houses or shops. They can determine which box would be best for which building by comparison of size.

For example, that house is the largest, this is the smallest, that house is smaller than the one next to it. The children could then make pipe-cleaner figures of about the correct height to live in the houses or to work in the shops.

2 Scale

Early work could compare the lengths of pairs of objects and express a relationship between the measurements, for example, the red car is twice as long as the blue car; this teddy bear is twice as tall as that teddy bear. You could also look at a set of Russian dolls and identify which is about twice the height and which half the height of a selected doll.

3 Scale drawings

Although scale drawings can be introduced as half or a quarter of real measurements, sometimes it is easier to make the relationship between units, for example, for every metre on the classroom floor have one centimetre on the paper. In another experience, the children could measure a shape and for every centimetre on the shape use a metre to make an enlargement for a display.

Gradually children should learn to interpret any scale factor, take the measurements of drawings and calculate the true measurements. They should also be able to make a scale drawing to a specified scale when given the real measurements.

3 Mapwork

Maps of different scales which show the local area can lead to finding true distances. This is particularly valuable when the children are trying to gain a reference for a kilometre and for a mile.

What are appropriate resources for teaching scale?

Model cars, model animals and representations of people allow children to see all the true measurements reduced by the same scale factor to produce a miniature which looks the same as the real item.

Advertisement material can often supply a scaled enlargement of an item like a chocolate bar or a drinks can.

What are possible contexts through which scale might be taught?

Any context where maps are involved, for example, the journey, the holiday, the country of . . . , a voyage around the world, travel by air, and the River Severn. In some scientific, technological and construction projects, the making of a scale model could be involved. In environmental projects where buildings, doors, street furniture, and other features are studied, the children could produce written work illustrated with scale drawings.

If models are collected in a project, for example, about transport, the farm or a battle, the children could check whether all the models are to the same scale.

How might scale be assessed?

Assess *orally* by asking children to:
- estimate the scale of a model soldier
- estimate the scale of an enlargement of a photograph
- explain what scale means in their own words.

Assess *practically* by asking children to:
- choose boxes to represent this classroom and those on either side, and then to explain their choice
- make a scale model of a classroom table and chair to the scale 1:10.

Assess *in written form* by asking children to:
- draw a lamp-post to scale, when given a true length, such as 10 metres and a scale factor such as 1:100
- draw a tree to the same scale as the lamp-post (whose true height is 10 metres) and then find the true height of the tree
- calculate the true lengths of the paths in a garden plan with the scale factor 1:50.

Assess through *problem solving* by asking children to:
- make a pipe-cleaner scaled figure of a friend, stating what scale they used.

What are common difficulties which children encounter and how might these be overcome?

Multiplying and dividing – It can be confusing for some children to know when to multiply and when to divide in scale calculations. This usually means that there is misunderstanding, or no understanding, about the scale relationship. The children would require more simple and practical examples to help gain the required concepts.

FIVE

Shape, position and movement

TWO-DIMENSIONAL SHAPES: SYMMETRY AND TESSELLATION

What does 2D shape mean?

Two-dimensional shapes are those which can be drawn on a surface, or plane, such as paper. They have length, breadth and area dimensions. In primary school, to allow children to handle 2D shapes, plastic or card representations, so-called 'flat' shapes, are used. The dimension of 'thickness' (height) is ignored in the flat shapes.

What are real life examples of 2D shapes?

There are many 'flat' shapes in the environment, but examples of 2D shapes are limited. Designs on materials, flooring and wallpaper often incorporate 2D shapes. Architectural drawings are based on 2D shapes.

What is the key vocabulary for 2D shapes and what does each word mean?

Geometry – the study of space which involves solid objects as well as properties and relations of constructible plane figures or shapes.
Figure – an arrangement of points, lines, curves and surfaces which make up a geometric shape. (It is interesting to note that the term figure is used to mean shape and also used to mean numeral.)
Point – a location in space, it has no length, breadth or thickness.
Line or line segment – can be thought of as a collection of points or as the shortest distance between two points. A line has length but no thickness.
Curve or arc – a drawing which starts and finishes without the drawing instrument being lifted from the surface.
Polygons – closed figures in a plane, constructed using three or more straight lines.
Triangles – polygons with three straight sides.
Quadrilaterals – polygons with four straight sides.
Pentagons – polygons with five straight sides.
Hexagon, heptagon, octagon, nonagon, decagon – . . .polygons with six, seven, eight, nine, ten . . . straight sides.
Regular – polygons which have all their sides equal and all their angles equal.
Similar – when two plane figures have corresponding angles equal, and so corresponding pairs of sides in proportion.
Congruent – plane figures are congruent when they are identical in size and shape such that they could be superimposed.
Symmetric – a plane figure identical with its own reflection in an axis or centre of symmetry

so that pairs of points and lines are identically placed.

Tessellation – covering a surface with identical shapes like a paving.

Some terms used for the properties of 2D shapes with primary children include:

Edge or side – the lines which form the boundary of the shape.

Corner or angle – where two lines which form part of the boundary of a 2D shape meet.

Diagonal – a straight line which joins any two non-adjacent corners to each other within a shape.

Perimeter – the boundary of a 2D shape.

What is the historical background of 2D shape?

The name geometry arises from Greek words which mean 'earth measurement'. People were required to survey the land and they did so by laying out shapes, each of which they could measure. It was the Greek mathematician Euclid who gathered together a number of self-evident facts, or axioms, about geometric shapes.

What is the value of teaching 2D shape?

Shape and size are part of the environment about which people want to communicate. It is useful for everyone to have the same conceptual background and vocabulary.

What are possible key steps in development for the learner?

1 An awareness of 2D shape
Children can be given gummed shapes and asked to select those they think appropriate to represent a house, a tree, the sun, a lorry, a head, a person and so on. They are able to extract the basic outline shape of objects around

them. Initially they have difficulty thinking in terms of three dimensions so they are happy to ignore the fact that often they are representing 3D objects as two-dimensional.

2 Descriptive vocabulary
The children can learn initially to describe 2D shapes in terms of their properties, for example, the rectangle could be described as 'four straight sides and four corners'. Young children like the unusual sound of a name like 'rhombus', so although it is unlikely to be meaningful for them, don't avoid such words. However, there is no need in the early stages to make the main aim of lessons the naming of shapes.

3 Sorting shapes
Flat shapes can be used to explore relationships. For example, children could find:
- all the shapes with only straight sides
- all the shapes with four sides
- all the shapes with more than four corners.

4 Symmetry about a line
Symmetry should be used as a means of finding out the properties of shape. However, the children require to know what symmetry means before they can use this concept.

Young children usually make their pictures symmetrical in that they assume a middle vertical axis and balance each item in the picture on one side of the axis by the same item on the other side. Here is an example:

Children could explore this symmetry or balance in activities such as placing pegs on a pegboard where an elastic band is positioned as a line of symmetry. After some individual practice, the children could work in pairs where one child is the leader placing the pegs and the other has the challenge of placing pegs of the appropriate colour in symmetric positions.

The children can make symmetric paint designs by placing a blob of thickish paint on paper and then folding the paper through the blob, pressing on the folded surface to spread out the paint and then opening the paper to see the symmetric design. The children can also try to cut out symmetric dolls, objects and patterns using scissors and a folded sheet of paper.

5 Bilateral symmetry of geometric shapes
The children could investigate shapes to find if they have an axis or axes of symmetry. They could then explore a shape to find out which sides and angles are equal by using the symmetrical property. For example, they would discover that a kite has one axis of symmetry which is a diagonal. Folding a paper shape along the diagonal line of symmetry, the children can find two pairs of sides and one pair of angles equal to each other.

6 Sorting by bilateral symmetry
The children could sort shapes according to their symmetry and discover, for example, that a rhombus and a square can be defined as kites as they all have one diagonal axis of symmetry.

7 Rotational symmetry
The children can investigate designs, such as those used as car manufacturers' badges, to find out how many times a tracing can be superimposed on the original through turning. They can then investigate geometric shapes to find how many times each fits into its own outline. All shapes fit once into their own outline. These shapes are regarded as of 'order 1' which means they do not have rotational

symmetry. Shapes with 'order 2' or more have rotational symmetry.

8 Using rotational symmetry to find properties
Rotational symmetry may be used to check the properties established by folding shapes, for example, the four equal sides and the four equal angles of the square will be found from folding or turning. However, rotational symmetry is particularly useful for shapes such as the parallelogram which has no bilateral symmetry. By turning the parallelogram, the opposite pairs of equal sides and the equal corresponding angles can be found.

9 Tessellations
The children can create tessellations by laying out a set of identical shapes without leaving any spaces. They can identify the shape which is tessellated to form different grids, for example, the equilateral triangle, the square and the regular hexagon. They can show tessellations of other shapes on such grids using different colours.

10 Tessellations to establish further properties
Symmetry can be used to establish equality of sides and angles in geometric shapes. Tessellations can be used to confirm the equality of sides and to find the size of angles, for example, all four equal angles of a square can meet at a single point. This means the sum of these angles is 360°, so each angle is 90°. Tessellations can also show lines which are always the same distance from each other, that is, parallel lines. This allows the opposite sides of a parallelogram to be stated as parallel.

11 Wallpaper patterns
Some children might enjoy identifying the part of the pattern which is the 'tile' used repeatedly to form the pattern on a wallpaper. This can be based on shapes such as the triangle, the square and the rectangle. Mark points which might be corners of a shape, join these and see what shape is produced. In patterns of geometric

shapes, the repeated 'tile' is often a composite shape.

12 Exploring movement
Some teachers may wish pupils to explore, possibly using a computer program, how shapes can be enlarged and reduced and how this does or does not affect equality of sides and angles. The children may also investigate what happens when a shape is moved across the plane (translation), reflected symmetrically and rotated symmetrically.

What are appropriate resources for teaching 2D shape?

Sets of 'flat' polygons, geostrips and paper fasteners to form the outlines of shapes, a geoboard to make shapes with elastic bands, pegboard and pegs, isometric and square grid paper, wallpaper (especially with geometric designs), rulers, compasses, set squares, scissors, mirrors and paper.

What are possible contexts through which 2D shape might be taught?

Art and design contexts such as making curtains and wallpaper, a study of the drawings of Escher, making a clothes pattern or an outfit all involve the children in interesting 2D shape work. The human body is like many of Nature's creations, an attempt to be symmetrical which is not achieved. Through a theme like 'Nature' this non-symmetry could be explored for humans, animals and plants. Children could contrast artificial human figures, flowers and leaves – the more symmetrical these are, the more unnatural they look.

How might 2D shape be assessed?

Assess *orally* by asking children to:
- select a flat shape from a bag and describe its properties

- give clues so that a classmate can guess which 2D shape is being thought about.

Assess *practically* by asking children to:
- find out the properties of a shape such as an equilateral triangle by folding it
- find out why Scott says a square could be called a rhombus and Ann says a square is a rectangle
- find out which of the following shapes tessellate: a rhombus, a regular pentagon and a kite
- sort some shapes into those with a diagonal axis of symmetry and those with a non-diagonal axis of symmetry.

Assess *in written form* by asking children to:
- draw a 2D shape with seven straight sides
- construct an equilateral triangle
- write instructions to draw a rhombus.

Assess through *problem solving* by asking children to:
- find the size of each angle in a regular hexagon without using a protractor
- identify the properties of an isosceles trapezium
- devise a checklist to use for properties of 2D shapes
- create a rotational symmetric design for a school badge.

What are common difficulties which children encounter and how might these be overcome?

Vocabulary – Some children will find the names difficult to remember. This may well stem from the names simply being labels to the child without any understanding of the properties which can be associated with each name. Some framework as to why certain shapes have such a name may be useful, for example, because of the number of sides: for example, three straight sides is a triangle, four straight sides is a quadrilateral, five straight sides is a pentagon. Children may need lots of practical and oral work where they describe shapes and make the shape words part of their vocabulary.

Understanding instructions, whether oral or written, can sometimes be difficult for some children if a number of mathematical terms are involved, for example, 'If the rhombus is folded along the diagonal axis of symmetry, you can identify the equal sides and angles'. Teachers should try to use simple sentences where children have to grasp only one or two terms at a time.

TRIANGLES

What does triangle mean?

A triangle is a two-dimensional shape with three straight sides.

What are real life examples of triangles?

The triangle provides strength and rigidity to constructions like bridges, cranes, gates, pylons, furniture, shelves and many other items.

What is the key vocabulary for triangles and what does each word mean?

Triangle – 'tri' is a prefix for three, so the word describes the shape as three angles.
Similar triangles – they have angles of the same size with sides which are in proportion.

The following terms are used to describe the properties of triangles:
Scalene – all three sides are of a different length.
Isosceles – two sides are the same length.
Equilateral – three sides are the same length.
Right-angled – the largest angle in the triangle is 90°.
Obtuse-angled – the largest angle in the triangle is greater than 90° but less than 180°.
Acute-angled – the largest angle in the triangle is less than 90°.

What is the historical background of triangles?

Triangles were particularly useful to ancient civilisations. The Egyptians found out that a triangle with sides of 3, 4 and 5 units was right-angled so they knotted a rope to form such lengths and then used this to construct right-angled corners when measuring land and buildings. Pythagoras in the sixth century BC established that the square on the largest side of a right-angled triangle equalled the sum of the squares on the other two sides. This knowledge allowed people to calculate the length of one of the sides of a right-angled triangle if the other two lengths were known.

Thales demonstrated that the ratio of lengths of any two corresponding sides in similar triangles is always the same. He used this to find the height of the Great Pyramid in Egypt.

To find the area of a plot of land with straight sides, divide it into triangles and find the area of each of these. This method has been used for many centuries.

What is the value of teaching triangles?

It is a common shape in the man-made environment and so to be able to communicate about triangles is useful.

What are possible key steps in development for the learner?

1 Recognising triangles

Children could be asked to sort triangles from a selection of 2D shapes. It is beneficial to include some shapes with curved edges, for example, a sector of a circle, so that the children realise that triangles have three straight sides. Some children do not recognise a triangle if it is not in a particular orientation so it is useful to present, and ask the children to show, triangles in a variety of orientations.

Children also should see triangles in a variety of shapes and sizes so that the name is not attached to any one type but generalised to all shapes with three straight sides.

2 Making triangles

The children can make triangles using: plastic strips and paper fasteners, a nailboard and elastic bands. They can draw triangles on: isometric and square grid paper, and the computer screen using an appropriate program.

3 The properties of triangles

The children are likely to be aware that there are different types of triangles so they could be asked to investigate in what ways they are different. They could find lines of symmetry where these exist. This should lead them to find that sides can be all equal in length, that two can be of equal length and that they can all be of different lengths. They can also identify that all three angles can be equal, two angles can be equal and that all angles can be of a different size. The children could also be asked to name the different types of angles found in each triangle and state the name of the largest angle.

4 Naming triangles

A triangle usually has two parts to its name. One part describes the sides – scalene, isosceles, equilateral – and the other names the largest angle – right-angled, obtuse-angled, acute-angled. The order of the parts does not matter.

This is a right-angled isosceles or isosceles right-angled triangle.

The children should find out that all triangles have two acute angles and it is only the third, the largest angle, which is different. They should also discover that all equilateral triangles have three equal acute angles and so they are rarely called acute-angled equilateral triangles. Some children will discover for themselves that there are seven different types of triangles:

- acute-angled scalene
- acute-angled isosceles
- right-angled scalene
- right-angled isosceles
- obtuse-angled scalene
- obtuse-angled isosceles
- equilateral.

The most difficult to recognise and to make or draw is the acute-angled scalene as this can be very like the acute-angled isosceles triangle.

5 The strength of the triangle

The children could investigate why the triangle is regarded as a 'strong' shape. If they make several 2D shapes using geostrips and paper fasteners, they should find that shapes with more than three sides are not rigid. To make four-, five-, and six- sided shapes rigid, additional strips need to be added as diagonals which form triangles. The children could identify the use of triangular features around the classroom and school, for example, shelf brackets and table corner pieces.

6 Forming other shapes with triangles

The children could also carry out investigations where pairs of congruent triangles, for example, obtuse-angled scalene triangles or acute-angled isosceles triangles, are placed side to side to make quadrilaterals.

A set of congruent triangles can be used

where two, three or more are combined to make shapes, for example, using equilateral triangles, a rhombus can be formed from two, an isosceles trapezium from three and five, a parallelogram from four, and a hexagon from four and from six.

What are appropriate resources for teaching triangles?

'Flat' plastic/card shapes, nailboards and elastic bands, geostrips and paper fasteners, scissors and paper, square and triangular grid paper.

What are possible contexts through which triangles might be taught?

Bridges, constructions, or making models could be used as a context for integrated work. There are many good computer programs which could also provide a motivating setting.

How might triangles be assessed?

Assess *orally* by asking children to:
- name triangles
- to describe a specific triangle.

Assess *practically* by asking children to:
- make three different types of triangles on the nailboard
- make three different shapes using two congruent right-angled isosceles triangles.

Assess *in written form* by asking children to:
- make five shapes by colouring a different number of triangles on isometric dot paper and then name their shapes.

Assess through *problem solving* by asking children to:
- find the sum of the angles of a triangle and describe how they did it
- find if all types of triangle tessellate and why this should be so.

What are common difficulties which children encounter and how might these be overcome?

Vocabulary – this requires to be developed over time. The children should learn to understand, use, read and write each of the terms, and to do so in a meaningful context. If the names are taught simply as a matching label for a drawing, the children are liable to confuse them.

QUADRILATERALS

What does quadrilateral mean?

A quadrilateral is a two-dimensional figure with four straight sides. It is a polygon.

What are real life examples of quadrilaterals?

The most usual quadrilaterals in the man-made environment are the rectangle and the square. These shapes are used for windows and doors in buildings.

What is the key vocabulary for quadrilaterals and what does each word mean?

Like all the 2D shapes, the terms 'side' and 'corner' are used to describe the shapes. Sides are referred to as straight, equal, unequal, opposite, adjacent and parallel. Parallel lines or sides are those which are always the same distance apart. As in the triangles, the corners (or vertices) of the quadrilaterals are described by the size of the angle, that is, right-angled, acute-angled and obtuse-angled. In quadrilaterals, it is also possible to have a corner which is a reflex angle which is greater than 180°.

Unlike triangles, quadrilaterals are not named by the equality of their sides and the size of the largest angle. There are seven types of quadrilaterals:

Parallelogram – has no line of bilateral symmetry, has rotational symmetry of order 2, has two pairs of opposite equal parallel sides and two pairs of opposite equal angles.

Rectangle – has two mid-point lines of symmetry, rotational symmetry of order 2, opposite sides equal and parallel and four right-angles.

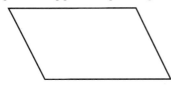

Kite – has one diagonal line of symmetry, two pairs of adjacent equal sides and opposite angles equal; there is a V-bomber or concave kite with the same properties and a reflex angle.

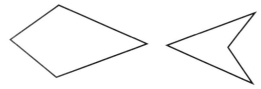

Rhombus – is a special kite with two diagonal lines of symmetry, rotational symmetry of order 2, four equal sides, two pairs of opposite equal angles.

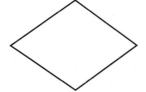

Square – is a special rectangle and a special rhombus as it has two mid-point and two diagonal lines of symmetry, rotational symmetry of order 4, four equal sides and four right angles.

Trapezium – no lines of symmetry, one pair of opposite parallel sides. There is also an isosceles trapezium which has one mid-point line of symmetry and two opposite equal sides.

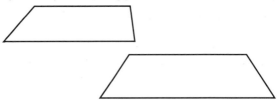

Quadrilateral – no line of symmetry and no special properties.

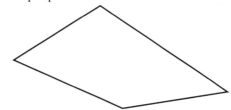

What is the historical background of quadrilaterals?

Geometric shapes were used in astrology to indicate the relative positions of two heavenly bodies with respect to the observer.

The golden section is when a line is divided so that the ratio of the shorter part to the longer part is the same as the ratio of the longer part to the whole. A golden rectangle is when the ratio of the difference between the side lengths equals the ratio of the shorter side to the longer side. To the Greeks the proportions of such a rectangle were uniquely pleasing to the eye and they used it in many of their buildings and works of art.

What is the value of teaching quadrilaterals?

The different quadrilaterals are used frequently in man-made products. It is useful for children to be able to recognise, draw and communicate about them.

What are possible key steps in development for the learner?

1 Recognising rectangles and squares
Children will first meet these shapes as 'four-sided flat shapes' rather than as quadrilaterals. The children could carry out activities such as:
- finding shapes which are the same by matching
- discussing the four sides and four corners of the 'window' or rectangle shape
- sorting shapes to find rectangles
- sorting 2D shapes to find all those with four sides
- discussing squares.

2 Making four-sided shapes
The children can make four-sided shapes using: plastic strips and paper fasteners, a nailboard and elastic bands. They can draw four-sided shapes on isometric and square grid paper, and the computer screen using an appropriate program.
Through making shapes with plastic strips or straws, the children could discover that:
- a rectangle pushed sideways becomes a parallelogram
- a square pushed sideways becomes a rhombus
- the corner between the shorter sides of a kite can be pushed inwards to give a concave or V-bomber kite.

Through making shapes on a nailboard, the children can investigate moving the elastic band from one corner to another.

- a rhombus into a kite

- a kite into a V-bomber kite

- a square into a trapezium

- a rectangle into a quadrilateral

Using either shapes on the nailboard or paper shapes, the children can divide one shape into others, for example:
- a parallelogram into a rectangle and two scalene right-angled triangles
- a kite into two acute-angled isosceles triangles
- a square into two isosceles right-angled triangles.

3 The properties of quadrilaterals
The children will learn that all shapes with four straight sides are called quadrilaterals. Through folding each quadrilateral the children can discover lines of symmetry and then identify equal sides and angles. Through rotating each

shape within its own outline, they can find if the shape has rotational symmetry and so confirm some equal sides and angles as well as identifying others.

It is only when carrying out tessellations with sets of the same shape that sides can be identified as parallel. The sizes of angles may be calculated at this time too, for example:

- all four angles of a quadrilateral have a sum of 360°
- angles of a square are 90°
- adjacent angles of a parallelogram add to 180°.

4 Relating the shapes to each other
Through sorting, especially for lines of symmetry, the children can find that:

- squares, rhombi and kites all have one diagonal line of symmetry – and so all are kites
- squares and rhombi have two diagonal lines of symmetry – and so both are rhombi
- squares and rectangles have two lines of mid-point symmetry – and so both are rectangles
- squares, rectangles and parallelograms are all parallelograms because their opposite sides are parallel.

5 Constructing quadrilaterals
Squares, rectangles, parallelograms and trapezia can be constructed using a ruler and a set square or a protractor. Rhombi and kites may be constructed using a ruler and a pair of compasses. It is useful to ask children to devise instructions to construct a specific shape.

What are appropriate resources for teaching quadrilaterals?

Children could recognise the different quadrilaterals in their environment, for example, the classroom, the school, and the nearby buildings, as well as in pictures of towns, buildings and objects. They should focus on 2D shapes and the faces of 3D shapes.

Plastic strips joined by brass paper fasteners (geostrips), straws joined by pipe-cleaners or Plasticine lumps can be used to form outlines of shapes. Quadrilaterals can also be made on a nailboard using elastic bands.

Sets of congruent plastic, card or paper shapes are ideal for tessellations. Paper shapes are also used for activities involving bilateral symmetry, while a plastic or card shape can have an outline drawn around it and be used for rotational symmetry investigations. Isometric and square grid paper are useful.

What are possible contexts through which quadrilaterals might be taught?

'Buildings' is an ideal theme to investigate as quadrilaterals are used for flooring, wall and even ceiling decoration, as well as forming wall, window and door shapes.

How might quadrilaterals be assessed?

Assess *orally* by asking children to:
- name different quadrilaterals
- describe a specific quadrilateral.

Assess *practically* by asking children to:
- make three different sizes of square on the nailboard
- make three different rectangles each with the same area on a nailboard
- identify equal sides and equal angles through folding a kite.

Assess *in written form* by asking children to:
- draw a rectangle, a square and a right-angled triangle, each with the area of 25 cm^2

Assess through *problem solving* by asking children to:
- find the sum of the angles of a quadrilateral and describe how they did it
- find if all types of quadrilaterals tessellate and why this should be so
- draw a parallelogram with an area of 30 cm^2 and describe how they did it.

What are common difficulties which children encounter and how might these be overcome?

Vocabulary – Some children find it difficult to say and remember the different names. Names should be introduced gradually and each should be related to a specific characteristic of the shape to help the children remember the name and the shape. The fact that a shape can have several names is puzzling for some children and it should be explained why a shape like a square can also be a rectangle, and a parallelogram, and a rhombus, and a kite! Snap cards and/or dominoes could be made to give the children practice in matching the most appropriate name to illustrated shapes.

Rote facts – The children should be encouraged to find properties of a shape from its symmetry rather than by trying to remember these by memory. A shape name can be associated with a specific product to aid recall.

CIRCLES

What does circle mean?

A circle is a two-dimensional shape which has one curved side. Every point on the curved side is the same distance from a fixed point which is called the centre of the circle.

What are real life examples of circles?

Many objects are made in a circular form, for example, a window, a mirror, a clock face.

What is the key vocabulary for circles and what does each word mean?

There are names which specifically describe parts of a circle, for example:
Circumference – the curved side.
Radius – any straight line drawn from the circumference to the centre of a circle.
Diameter – any straight line joining two points on the circumference which passes through the centre, twice the length of the radius.
Chord – any straight line joining two points on the circumference which does not pass through the centre.
Semi-circle – bounded by the diameter and half of the circumference with an area of half of the circle.
Sector – bounded by two radii and part of the circumference.
Segment – bounded by part of the circumference and a chord.

What is the historical background of circles?

Circles of standing stones like those at Stonehenge and in Orkney at Stenness illustrate the religious significance given in the past to this shape.

What is the value of teaching circles?

The circle has special properties which make it interesting for pupils to investigate.

What are possible steps in development for the learner?

1 Recognising a circle
Young children when sorting 'flat' representations of 2D shapes recognise the circle because it does not have any corners or straight sides. The one curved side is associated with rolling and wheels. This can cause

confusion between the 2D circle and the 3D cylinder.

2 Circles as faces of 3D shapes
Gradually the difference between 2D and 3D shapes is understood and the circle identified as a face of a cone and the two congruent faces of the cylinder.

3 Length measurements of a circle
The sides of shapes like a triangle, a square and a rectangle are usually measured with a ruler. The one curved side of the circle presents a greater challenge as the tape-measure, or string, needs to be used. The perimeter of 2D shapes is found by calculating the sum of the side lengths. For the circle, the perimeter and the length of the one side, the circumference, is the same.

If a circle is drawn with the centre marked, children can measure the diameter using a ruler. However, if challenged to find the diameter of a circular object, the children need to consider how to identify the 'broadest' part or the greatest length from circumference to circumference. This can be found by placing the object between two parallel straight edges such as books, then measuring the distance between the edges.

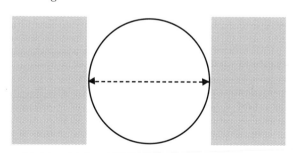

The radius can be found as half the measurement of the diameter.

4 Pi
Some children might be set the task of finding the relationship between the length of the circumference and the length of the diameter. They could measure then calculate to find

circumference ÷ diameter

for several objects. They should find that the circumference is about three times the length of the diameter. The children might be told that, the answer of all circles, when based on accurate measurements, gives a very special number which has been calculated to over 1011 million decimal places! It begins like this:
3·141 592 653 589 79 . . .

Because this number cannot be written down the Greek letter pi, written as π, is used to represent it. In calculations, an approximate value such as 3·14 is used. Pi is an example of an irrational number, that is a number which cannot be expressed as a ratio of two integers. The circumference can be expressed as πd or $2\pi r$.

5 Drawing circles
Pupils can explore drawing circles on the floor or table with a length of string, a drawing pin and a pencil. They can then investigate compasses as an instrument where the pencil can be 'set' at a specific length from the vertical point which will mark the centre. They can also explore drawing a circle on the computer screen with logo commands.

6 The area of a circle
The approximate area of a circle can be found by several methods, for example:
- by superimposing the circle on a square grid then counting the whole squares, the half squares and those greater than half which lie within the circumference
- by drawing a square within the circle and a square outside the circle, both touching the circumference, and allocating a value midway between the areas of the squares to the circle
- by cutting a gummed circle into equal sectors and repositioning these to form a rectangle,

the area of the circle is $\frac{1}{2}C \times r$, or $\frac{1}{2}(2\pi r)r$, or πr^2.

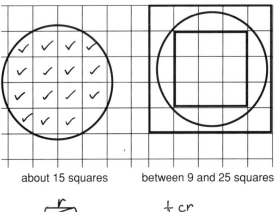

about 15 squares   between 9 and 25 squares

$$\frac{1}{2}cr$$
$$=\frac{1}{2}\pi dr$$
$$=\frac{1}{2}\cdot 2\pi rr$$
$$=\pi r^2$$

7 Other investigations
Pupils can investigate drawing circle patterns – overlapping circles, concurrent circles and others. They can form circular patterns by drawing straight lines from points on two intersecting axes. Such patterns can be sewn and are known as curve stitching.

Pupils can explore forming the perimeter of several shapes – triangle, square, rectangle and circle – with a pipe-cleaner, to identify which shape of perimeter encloses the greatest area. Similarly, pupils can use congruent paper rectangles to find which shape encloses the greatest volume. They can make a triangular prism, a cuboid and a cylinder, calculating the volume as the area of the base times the height in each instance.

What are appropriate resources for teaching circles?

Circular objects like mugs, jars and plates. Gummed circles, centimetre squared paper and an acetate grid, centimetre tape, ruler, scissors and a calculator.

What are possible contexts through which circles might be taught?

Making circle patterns using compasses, investigating perimeters, and investigating volumes are three examples of mathematics based contexts.

How might circles by assessed?

Assess *orally* by asking children to:
- describe a circle
- give instructions to a friend to draw a circle with a radius of 6 cm
- calculate the circumference of a circle with a radius of 10 cm
- calculate the area of a circle with a radius of 10 cm.

Assess *practically* by asking children to:
- sort all the circles from a mixed set of 'flat' shapes
- draw a circle with a diameter of 16 cm
- make a semi-circle where the radius is 10 cm

Assess *in written form* by asking children to:
- draw diagrams to illustrate the difference between a sector and a segment
- follow instructions such as 'Draw two circles with the same centre, one should have a radius of 6 cm and the other 8 cm' or 'Draw a circle with a radius of 8 cm. Choose any point on the circumference and the same radius to draw an arc within the circle and then make other arcs (putting the compass point where each arc touches the circumference) to complete the design.'

Assess *through problem solving* by asking children to:
- design a wheel cap which has rotational symmetry
- find out why the cylindrical can is used so often to package foods

- how to draw an ellipse using a piece of string, drawing pins and a pencil.

What are common difficulties which children encounter and how might these be overcome?

Vocabulary – such are sector and segment might be confused but this should be overcome if the children make their own book or poster of 'circle words'.

Formulae – if the pupils try to learn that $C = 2\pi r$ and $A = \pi r^2$ without understnding, they are likely to confuse them. Working on a structured task which leads them to 'discover' each formula should help to give them understanding.

3D SHAPES

What does 3D shape mean?

Three-dimensional shapes have a length, a breadth and a height. They can be called polyhedra which means 'many faces'.

What are real life examples of 3D shapes?

Buildings and many objects are examples of three-dimensional shapes, for example, a block of flats, a tin of soup, a cricket ball. Some buildings such as churches are often several shapes joined together whereas an object like a lampshade is a part or a frustrum of the 3D shape called a cone.

What is the key vocabulary for 3D shapes and what does each word mean?

To describe the properties of 3D shapes the following terms should be used:

Face – a surface of the shape which can either be flat or curved.

Edge – two faces meet at an edge which can be either straight or curved.

Vertex – two or more edges meet at a vertex or corner (the plural of vertex is vertices).

Many polyhedra can be classified under two 'family' names:

Pyramids – 3D shapes which can be cut into a series of similar slices (the slices are the same shape with the edges in the same ratio), for example, a cone could be cut intd a series of circle slices, with the largest circle slice at the base and the smallest at the vertex or apex.

Prisms – 3D shapes which can be cut into identical or congruent slices, for example, a cube could be cut into identical square slices.

NB Not all shapes belong to these two families, for example, the sphere.

Five polyhedra are regarded as very special. Sometimes they are called the **Platonic solids** after Plato who discovered them, but they are also known as the regular 3D shapes. Regular means that all faces are identical, all edges are of the same length and all angles are the same size. A regular shape, such as the cube, can be turned over onto another face and it still looks exactly the same.

The most common geometric 3D shapes are:

Cube – a regular prism with six flat congruent square faces, twelve straight equal edges and eight vertices.

Cuboid – a prism with six flat rectangular faces which are arranged as opposite congruent pairs, twelve straight edges which form three sets of four of equal length (sometimes eight are of the same length) and eight vertices.

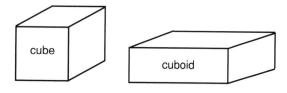

Triangular prism – five flat faces with two opposite congruent triangular faces and three rectangular faces, two or three of which may be congruent, nine straight edges of which three must be the same length, and six vertices.

Pentagonal prism – seven flat faces with two opposite congruent pentagonal faces and five rectangular faces, fifteen straight edges of which five must be the same length, and ten vertices.

Cylinder – a prism with three faces of which two are congruent flat circles and one is curved, two equal curved edges and no vertices.

Other prisms include the **hexagonal prism**, the **heptagonal prism**, and the **octagonal prism** – these are called after the name of the 2D shape which forms the two opposite congruent faces (and the slices).

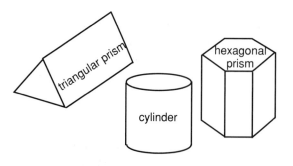

If the base face (that is, the shape of the similar slices) can have a circle drawn through all its vertices, and if the perpendicular from the apex passes through the centre of this circle, then the pyramid is said to be a **right pyramid**.

The following descriptions refer to right pyramids:

Cone – a pyramid with two faces, one a flat circle and the other curved, one curved edge and one vertex or apex where no edges meet.

Triangular pyramid – a pyramid with four triangular faces, six straight edges of which three are equal, and four vertices.

Square pyramid – a pyramid with five flat faces of which one is a square and the others are congruent triangles, eight straight edges of which there are two sets of four equal lengths, and five vertices.

Pentagonal pyramid – a pyramid with six flat faces, one of which is a pentagon and the others triangles, ten straight edges and six vertices. Other pyramids include the **hexagonal pyramid**, the **heptagonal pyramid**, and the **octagonal pyramid**.

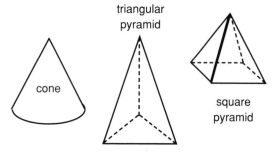

triangular pyramid

cone

square pyramid

The cube is a regular shape. The other four regular shapes are:

Regular tetrahedron – a pyramid with four flat congruent equilateral triangle faces, six straight equal edges and four vertices.

Regular octahedron – eight flat congruent equilateral triangle faces, twelve straight equal edges and six vertices.

Regular dodecahedron – twelve ('do' means two and 'deca' means ten) flat congruent pentagon faces, thirty straight equal edges and twenty vertices.

Regular isoscahedron – twenty flat congruent equilateral faces, thirty straight equal edges and twelve vertices.

Another polyhedron is:

Sphere – one curved face with no edges or vertices.

Polyhedra can be named in different ways, for example, a cuboid is the common name for a shape which can also be called a rectangular prism or a hexahedron. If you don't know a name for any 3D shape it can always be called by the number of faces it has, for example, a tetrahedron (4 faces), a pentahedron (5 faces), a hexahedron (6 faces), a heptahedron (7 faces), an octahedron (8 faces) and so on.

A **net** is the arrangement of 2D shapes which can be used to construct a 3D shape.

What is the historical background of 3D shapes?

Plato, about 400 BC, gave the five regular polyhedra mystical significance. Later, about 1600, Kepler was led to the discovery of how planets moved by using facts about the Platonic solids in his calculations.

What is the value of teaching 3D shapes?

Teaching about 3D shape allows children to communicate unambiguously about shape. It also improves their perception of objects. It can also, in true Platonic tradition, challenge individuals to investigate relationships among shapes.

What are possible key steps in development for the learner?

1 Handling 3D shapes
Young children can explore shapes to find if they roll or slide, if they are easy to build with or not, and begin to link these properties with the shape's flat and 'round' faces. Some children will be able to describe the movement of rolling

shapes like the sphere, cone and cylinder, and a few might relate the different movements to the number of curved edges, that is, none, one and two.

2 Realising that 3D shapes have 2D faces
The names of 3D and 2D shapes are readily confused by young children. For example, a cube is referred to as a 'square' and a cylinder as a 'circle'. This is understandable, especially as we use 'flat' shapes which are really three-dimensional to represent 2D shapes. The children may already know some names but now they should become familiar with sphere, cylinder, cone, cube and cuboid. Some may also learn the names pyramid and triangular prism, especially as these shapes are often used for sweet packaging. Drawing around the faces of 3D shapes, cutting up cartons into individual faces, or painting the faces of 3D shapes to 'stamp' the shape of the face, should help children to begin to appreciate the relationship between 3D and 2D.

3 Faces, edges and vertices
The children need to understand that a face is a surface, an edge is where two surfaces meet and a corner is where edges meet. Faces can be identified by drawing a face on each surface. Faces can be counted by marking each face with a number. Edges can be marked with paint or chalk as they are counted. Corners can be topped with a blob of Plasticine as they are counted. The children can also begin to investigate the shape of the faces and identify which are identical.

4 Constructing 3D shapes
If the children are aware of the number and shapes of the faces, they can consider constructing 3D shapes. An ideal material is the plastic flat shapes which clip together. However, plastic or card shapes may be taped together to produce satisfactory results. As this work is carried out the children could be challenged to

find if 3D shapes can be made, for example, with a set of congruent equilateral triangles, with a set of congruent squares, with a set of regular pentagons and with a set of regular hexagons. Children should find it is possible to construct 3D shapes with all the 2D shapes mentioned, except the hexagons which fit together to form a flat surface, not a three-dimensional one.

Prisms and pyramids (except the cylinder and cone) can also be built with straws and pipe-cleaners or other construction materials producing 'skeleton' shapes. These are useful to focus on edges and vertices. The children can investigate equal and unequal edges.

5 Prisms and pyramids

With the experience of having made 3D shapes, the children should be able to classify those that come to a special corner or apex and those that don't. This gradually leads them to understand the nature of a prism and of a pyramid. Towers of congruent 'flat' shapes, for example triangles, squares, rectangles, pentagons, and hexagons, will provide the ideal explanation of prisms with their uniform cross-section. Towers of similar shapes will provide an acceptable model for the pyramids. The cylinder is often regarded as the unusual prism and the cone is certainly considered the unusual pyramid because it does not have any triangular faces.

When the children sort shapes into a set of prisms and a set of pyramids, they will find that there are no common members (the sets are disjoint). The pupils can also investigate the patterns produced by the numbers in each of the columns of faces, edges and corners, for example:

	faces	vertices	edges
triangular prism	5	6	9
rectangular prism	6	8	12
pentagonal prism	7	10	15
hexagonal prism	8	12	18

	faces	vertices	edges
triangular pyramid	4	4	6
rectangular pyramid	5	5	8
pentagonal pyramid	6	6	10
hexagonal pyramid	7	7	12

6 Nets

Children can make nets by cutting along some, but **not all**, edges of cartons to produce an arrangement of the faces where each is joined by at least one edge to another (flaps may need to be cut off or explained). Some children may like to investigate using either cartons or flat shapes joined together, how many different nets can be made for the same shape, for example, a cube, or a regular tetrahedron.

The children may also like to be challenged to make and/or draw the net of a cone and a cylinder. They are usually interested to find the curved faces may be produced by a sector of a circle and a rectangle.

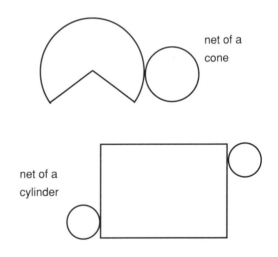

net of a cone

net of a cylinder

7 Another investigation

An interesting investigation for many pupils is to find how the number of faces, edges and corners for each shape relate to each other. The relationship: faces + vertices − edges = 2 was first discovered by Leonhard Euler, a prolific Swiss mathematician, in the late eighteenth century.

What are appropriate resources for teaching 3D shapes?

Building blocks, 'flat' plastic shapes, 3D shapes, cartons, Plasticine, straws, pipe-cleaners, card, scissors, 2 cm or 5 cm squared paper, glue.

What are possible contexts through which 3D shapes might be taught?

Model building, making gift boxes, decorations, studying houses and other buildings in the environment.

How might 3D shapes be assessed?

Assess *orally* by asking children to:
- count the number of faces on a given shape
- describe a shape
- give clues for friends to guess a hidden shape.

Assess *practically* by asking children to:
- construct a triangular prism from a given selection of plastic shapes
- make a net from a given carton

- identify the basic geometric shapes which have been used to make a model
- make two different nets of a cube using squared paper.

Assess *in written form* by asking children to:
- sketch the net of a cylinder
- design and sketch a model building based on geometric shapes.

Assess through *problem solving* by asking children to:
- make two different 3D shapes, each with three faces
- make eight different 3D shapes, each from four centimetre cubes.

What are common difficulties which children encounter and how might these be overcome?

Confusion between 3D and 2D names – Some children try to remember shape names by rote and become confused. They usually require more experience in sorting shapes where they have the opportunity to discuss the properties and the names of shapes.

LINES

What does line mean?

A line joins two points. Lines can be straight or curved. A straight line can be defined as a one-dimensional geometric figure with infinite length and no thickness. Some straight lines have special properties, for example: a diagonal, a line of symmetry, parallel lines.

What are real life examples of lines?

Railway lines are parallel. The crossbar on a rectangular gate is usually diagonal. Doors, windows and pictures are usually placed with their edges horizontal and vertical.

What is the key vocabulary for lines and what does each word mean?

Vertical or **perpendicular** – when a line is pointing in the direction of the Earth's centre it is said to be vertical; the vertical direction is determined by using a plumb line, that is, a weight suspended by a length of string.

Horizontal – when a line is at right angles to the vertical; in practical situations a spirit level can be used to determine if a surface or plane is horizontal.

Diagonal – a straight line joining any two vertices or corners which are not adjacent in a shape (2D and 3D).

Mid-point line – a mid-point is the point on a line that is equidistant from its end points; a line drawn from the mid-point of one side to the mid-point of an opposite side in a shape like a rectangle can be called a mid-point line.

Altitude – a line drawn from a vertex to a side of a polygon which is perpendicular to the side; used particularly to refer to any perpendicular from a vertex to the opposite side in a triangle.

Line of symmetry – the fold line around which pairs of points are identically placed.

Parallel lines – lines which remain a constant distance apart and so never meet or intersect.

What is the historical background of lines?

Used by the Greeks in their investigations into geometric shapes.

What is the value of teaching lines?

The vocabulary associated with lines allows clearer communication.

What are possible key steps in development for the learner?

1 Straight and curved

The children are likely to become aware of lines as edges and sides of shapes. They can find out that some of these lines are straight and some curved. Curved edges give the shape the ability to roll. The one curved edge of the cone rolls to trace a circle while the apex marks the centre. The two curved edges of the cylinder trace out parallel lines. The children should be shown how to draw a straight line using the edge of a ruler.

2 Lines of symmetry

When investigating two-dimensional shapes, children will find 'fold' lines which allow one half of a shape to be aligned with the other. These fold lines can be lines which join two vertices (a diagonal), one vertex and one line (an altitude), or two lines. Such lines are characteristic of different families of shapes, for example, kites have one diagonal line of symmetry, rectangles have two mid-point lines of symmetry, squares have two diagonal and two mid-point lines of symmetry.

3 Vertical and horizontal

Children can make a plumb line and then use this to find if the edges of doors and windows are vertical. They can use a spirit level to find out if the surfaces of window ledges, tables, and shelves are horizontal.

4 At right angles

A set square can be introduced to the children and used to identify right-angled corners. Lines can be constructed at right angles by using a ruler and a set square.

5 Parallel lines

These can be constructed by drawing along the two edges of a ruler. They can also be drawn by using a ruler and set square. First draw along the edge of the ruler, then mark a sequence of points the same distance from the line and at right angles to it using the set square. Join the points to make the parallel line.

Lines drawn parallel to the edges of a rectangular sheet of paper are usually regarded as horizontal and vertical and may be referred to as such. Pupils can identify parallel lines in tessellations made from shapes. They should also watch for examples of parallel lines in real life, for example, telephone and electric wires on pylons, tramlines, railway lines, path and road edges.

What are appropriate resources for teaching lines?

Ruler, set square, plumb line, string, Plasticine, spirit level, the environment, photographs of the environment.

What are possible contexts through which lines might be taught?

Railways, buildings, and model making are possible contexts for integrated work.

How might lines be assessed?

Assess *orally* by asking children to:
- name two horizontal and two vertical surfaces in the room
- identify sets of parallel lines in a photograph
- give instructions to draw parallel lines which are six centimetres apart.

Assess *practically* by asking children to:
- use a plumb line to find if the cupboard, the door and the bookcase are vertical
- find any lines of symmetry of each of the following three shapes: a rhombus, a parallelogram and an equilateral triangle.

Assess *in written form* by asking children to:
- make a design of parallel lines
- identify parallel lines in a design
- identify parallel lines in shapes.

Assess through *problem solving* by asking children to:
- make a spirit level using a marble
- draw a flower head design with three lines of symmetry.

What are common difficulties which children encounter and how might these be overcome?

Equipment – Some children are likely to need practice in using a ruler and set square before becoming efficient. One child was found to draw a line near the ruler rather than use it to guide the pencil point.

ANGLES, COMPASS POINTS AND BEARINGS

What are angles, compass points and bearings?

An angle is the amount of rotation of one line from another at a fixed point, for example:

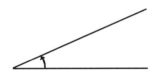

Angles are measured in degrees (°), with a full revolution being considered as 360°.

Directions can be communicated using the compass points of north, north-east, east, south-east, south, south-west, west, and north-west. A bearing is another means of communicating direction in which the amount of rotation is measured from:
- north or from south, indicating whether the direction of the turn is west or east, for

example: N 36° W (the angle is always between 0° and 90°)
- north and is always expressed as three figures, for example:

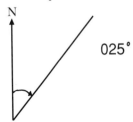

025°

What are real life examples of angles, compass points and bearings?

Edges of shapes meet at an angle. A shelf is usually at a right angle to the wall. You can raise your arm at an angle from your body. Games such as snooker, football and squash require a knowledge of angles.

Compass points or bearings can be used to

communicate directions for a walk in woods or moorland, for example, 'from the white gate, walk in the direction 065°'.

What is the key vocabulary and what does each word mean?

Angle – the amount of turning of one line from another at a point on both lines.
Degree – the unit of angular measurement, one degree is a three-hundred-and-sixtieth part of a full rotation of a line about one of its endpoints.
Rotate – to turn.

Angles are classified by the amount of rotation as follows:
Acute angle – less than 90°.
Right angle – 90°, often called a square corner or square angle; the lines forming the angle are perpendicular to each other.
Obtuse angle – greater than 90° and less than 180°.
Straight angle – 180°, the two arms of the angle appear as a straight line.
Reflex angle – greater than 180° and less than 360°.
Full revolution – a complete circle turn, measured as 360°.
Compass – an instrument with a magnetised needle which indicates direction.
Bearing – a form of communication of direction which combines the compass point north and the angle of rotation, and is expressed as the position of one object with reference to another.
Protractor – the instrument which measures the angle in degrees.

What is the historic background of angle, compass points and bearings?

Geometry can be thought of as the study of space. The name 'geometry' arises from the Greek words meaning 'earth measurement'. The Egyptians measured their land every year because of the flooding of the Nile and they developed rules involving measures and angles to enable records to be kept.

Aircraft navigation uses bearings as modern travel requires greater accuracy than the mariner's compass with its thirty-two points of direction (N, N by E, NNE, NE by N, NE, NE by E, ENE, E by N, E etc.).

Why should we teach angles, compass points and bearings?

These are part of communication about shapes and directions. A knowledge of angles is essential to engineers, surveyors, architects, builders and navigators.

What are possible key steps in development for the learner?

An angle can be thought of as the shape of a corner or as an amount of turning. Any development tends to focus on one of these concepts, then the other, and finally the integration of both.

1 Direction and turning
Development can begin with a focus on direction and turning, for example, the children could consider direction initially as right or left. When turning to the right or left, the amount of turn is usually expressed as a full turn, a half turn or a quarter turn. The use of computerised toys and logo programs could help the children to see the different amounts of turning and appreciate different directions. It is beneficial to let the children follow the toy so that they are always facing the same way, and making the same turns, as the toy.

2 Compass points
The direction of the turn can be linked to the compass points of north, south, west and east. These points should be related to environmental landmarks, for example, if we look out of the window towards the church, we

are looking towards the north. The direction of north can be marked on the classroom floor and in the playground. The children could experience making quarter and half turns to learn the other three main compass points and how they are related to north.

3 Corners

When looking at shapes, children notice that the corners vary. The square corner can become the reference with others found to be less than a square corner and greater than a square corner.

4 Turns as degrees

The amount of turn can be expressed in degrees. The children soon realise that a degree is a unit of measurement just like centimetres. It is important that they understand that they are measuring the amount of turning, whereas centimetres is the amount of length or distance. A full turn is the measure 360°, so a half turn is 180° and a quarter turn 90°. Such measures are perceptual rather than measured at this stage. The children can experience turning themselves, can see computer toys make these turns, and program instructions with such measurements on the computer screen.

5 Corners and angles

Two sides meet at a corner and the size of the corner is determined by the angle between the sides, in other words, the corner and the angle at the vertex are the same thing. Corners can now be expressed as a right angle, less than a right angle or an acute angle, and greater than a right angle or an obtuse angle.

6 Eight compass points

Comparison can be made between quarter turns, right angle turns and 90° to find that these are all the same amount of turn. The concept of a 'half right angle' could be related to the compass directions such as north-east which is half-way between north and east. The children may be surprised to find we do not use the expression east-south but south-east.

Convention is that the name north or south comes first for naming these points.

7 Properties of shapes

When describing shapes, angles are usually referred to by their type, for example:

The square has four right angles:

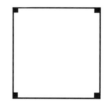

This is an acute-angled isosceles triangle:

8 Use of a protractor

There are two types of protractor. One is circular in shape and shows the full 360°, the other is a little larger than a semicircle and is marked 0° to 180° in the direction left to right, and 180° to 360° in the direction right to left. Children find the full circle version easier to use. The main intervals on the protractor scale are usually 10°, and pupils can measure angles to the nearest 10° in the initial stages. Later an accuracy of 5° may be used.

It is usually easier for children to measure angles than to draw them. The children should find out that the protractor is placed so that its centre is at the vertex of the angle. The line indicating 0° on the protractor is positioned on top of one arm of the angle to be measured, so that the 'turn' of the angle can be measured clockwise.

When drawing angles, one arm of the angle is represented by a straight line. One end of the line is aligned with the centre and the 0° line on the protractor and then a mark placed at the required point on the protractor scale. When the protractor is removed the mark can be linked to the end of the line to make the required angle.

9 Tessellations

Investigations of shapes which tile, that is, fit together over a surface without leaving any spaces, can lead to finding out the size of angles. For example, in a tiling of regular hexagons you find three equal angles positioned around a point or vertex, so it can be calculated that each is 360° ÷ 3 or 120°.

10 Constructing shapes

When making scale drawings or constructing shapes, pupils can draw angles with a protractor. They can investigate how to draw, for example, an equilateral triangle using a ruler and protractor and by using a ruler and a compass. Shapes can also be drawn on a computer screen where the pupil devises the instructions for lengths and angles.

11 Mapwork

Compass directions can be used to plan journeys on maps. Pupils learn to move from one position to another by stating the direction and the distance. Initially, examples should be planned so that the accuracy of direction involves only the eight compass points.

Pupils can be told that greater accuracy for direction can involve sixteen or thirty-two compass points, but that expressing direction as a number of degrees from north is now the more usual form used. Examples should involve finding directions on a map and following given directions. Map examples where the pupils find the directions can be constructed with circles at the different points of reference. A small protractor which fits into the circles can be supplied for these specific examples. (This format for examples is used in SPMG *Mathematics Stage 5* and *Heinemann Mathematics 7*.)

What are appropriate resources for teaching angles, compass points and bearings?

Local direction landmarks, a compass transparency to superimpose on a map, a compass, plans and maps, computer software, computerised toys, the logo turtle, a circular protractor, flat shapes, tessellation grids.

What are possible contexts through which angles, compass points and bearings might be taught?

Activities could involve following, or giving, directions when blindfolded; investigating or planning journeys. Bearings are best associated with orienteering and plane journeys on a map. Extended contexts of projects might include: blindness, treasure, the field trip, holidays, and a voyage.

How might you assess angles, compass points and bearings?

Assess *orally* and *practically* by asking children to:
- identify the compass direction of objects in the room when given north
- identify an object positioned in a given direction
- turn through a three-quarter turn from a given direction
- identify the direction faced after a half turn given the starting direction
- indicate the approximate direction of the bearing 100°, given north
- follow given directions from the classroom and identify their final destination
- follow directions on a map and identify a 'meeting point'.

Assess *in written form* by asking children to:
- list directions for someone to find hidden treasure
- make a scale drawing of the school perimeter
- design, and draw to the scale of 10 cm for each 1 m, the command room of a space ship.

Assess through *problem solving* by asking children to:
- plan and list instructions for an orienteering walk for the class
- plan, prepare clues and carry out a treasure hunt
- draw a map to show, with approximate directions and distances, the journey to school for a pupil who is coming to live at a given address.

What are common difficulties which children encounter and how might these be overcome?

Right from left – Many people confuse their right from their left and have to remember something like 'I write with this hand so it is my right hand'. Often east and west are interchanged because of a child trying to relate these directions to right and left. The children should be told that, when facing north, west and east are in the same order as in the word 'we'.

Size of angle – Some children mistakenly believe that the size of an angle depends on the length of the arms of the angle.

NETWORKS AND TOPOLOGY

What do networks and topology mean?

A network is a diagram where points are linked by lines. Topology is the study of properties of shapes which remain unchanged when the shape is transformed through bending, stretching, squashing or twisting. Two-dimensional topology is often thought of as 'rubber sheet geometry'.

In both networks and topology, distance, size and straightness are not important, whereas configuration is.

What are real life examples of networks and topology?

Roadways, railways, flight paths, telephone links are all examples of networks. Topology tends to feature in puzzles where the challenge is to separate loops of string without cutting or untying knots. Children could be shown fan belts where a twist allows the belt to have one continuous surface.

What is the key vocabulary for networks and topology and what does each word mean?

Configuration – spatial arrangement.
Node – a vertex or point of a network where two or more lines meet.
Arc – a line or continuous path of a network.
Source – a node with no entering arcs.
Terminal or sink – a node with no exiting arcs.
Transformation – in topology this involves an object being bent, stretched, compressed or twisted or any combination of these.

What is the historical background of networks and topology?

Topology is only about one hundred years old, although Euler and Gauss had provided a foundation earlier for Henri Poincaré and others to build the concepts of topology. Euler, the Swiss mathematician, devised formulae for networks in the eighteenth century. He devised a formula about vertices, edges or arcs and regions or faces: $V - E + F = 1$.

It is said that Euler became interested in networks because of the problem conceived by the people of the town of Konigsberg which involved islands and bridges. They wanted to know if it was possible to walk across all seven bridges without crossing any of them twice. Euler differentiated between even nodes (with an even number of arcs going to or from it) and odd nodes (with an odd number of arcs). The complete network can only be travelled once without going over any point twice if the number of odd vertices is exactly 0 or 2. In this drawing of the Konisberg Bridge problem you can find that there are four odd vertices so that it is impossible to travel over each bridge only once and yet pass over all of them.

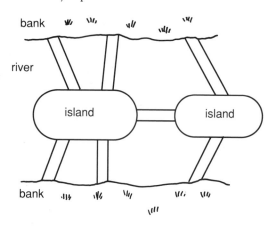

What is the value of teaching networks and topology?

Networks give rise to interesting problems which might be investigated. Rules and formulae are best left until the pupils are older.

What are possible key steps in development for the learner?

1 Using diagrams
Pupils could be given diagrammatic problems to solve for example: Find out if you can trace the following network without lifting the pencil

from the paper and without going over any lines twice.

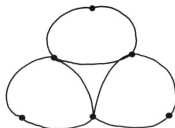

2 Drawing diagrams
For primary pupils network diagrams can be a means of expressing a problem and lead to a solution.

It is useful to provide simple maps which could be expressed as a network diagram.

What are appropriate resources and contexts for teaching networks and topology?

Paper, pencil, scissors, a belt, maps reference books and commercial network and topological puzzles could be useful. Maps provide a suitable context.

How might they be assessed?

Assess through *problem solving* by asking children problems such as: Find out if a postman who has to deliver letters to the streets represented in the network below can do so without travelling along the same street twice.

What are common difficulties which children encounter and how might these be overcome?

Some pupils might find it difficult to accept the abstract representation in a network diagram where distance and direction do not matter. Experience of more examples should help.

SIX

Information handling

PICTORIAL REPRESENTATION

What does pictorial representation mean?

Quantities can be displayed through pictures and diagrams. This can make them easier to compare and to find relationships between them. One strategy in mathematical problem solving is to draw a diagram as this often clarifies the problem to be solved and/or can suggest a method of solution. The best known type of mathematical diagram is a graph. This is dealt with under a separate heading. In this section the focus is on arrow diagrams, Venn diagrams, Carroll diagrams and tree diagrams.

What are real life examples of pictorial representation?

Posters, adverts and the media make use of pictorial representation to illustrate information.

What is the key vocabulary for pictorial representation and what does each word mean?

Data – facts.
Arrow diagram – a relationship diagram where elements of two sets are linked by arrows; all the arrows represent the same relationship, for example, 'is taller than', or 'has'. The element (letter, number or word, such as John) before

the tail of the arrow, begins a sentence in which the arrow supplies the relationship (such as, 'has as a pet') and the element (such as a dog) at the head of the arrow completes the sentence.

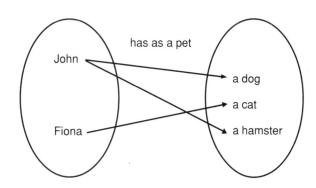

Tree diagram – a display of sorting into sets by following labelled pathways. Consider the set of numbers one to ten by directing each number along the correct path according to whether it is less than six and whether it is odd (you can do this using each number written on a slip of paper).
The boxes shown in the diagram on page 114 contain the following sets:
- numbers which are less than 6 and odd
- numbers which are less than 6 but not odd
- numbers which are not less than 6 but odd
- numbers which are neither less than 6 nor odd.

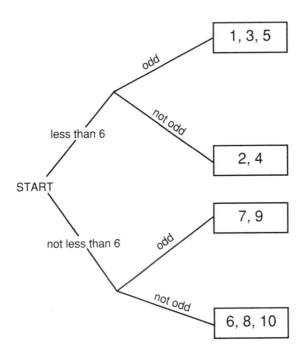

Carroll diagram – a display of sorting through the use of 'boxes' without the paths used in a tree diagram. The boxes are labelled according to the column and the row.

	less than 6	not less than 6
odd	1, 3, 5	7, 9
not odd	2, 4	6, 8, 10

Venn diagrams – sets are represented by regions outlined by a boundary such as a rectangle, a loop or circle. Sometimes the boundaries overlap, for example in the diagram below, the main or universal set which is labelled E is represented by a rectangle and two sets, labelled X and Y, are shown by circles.

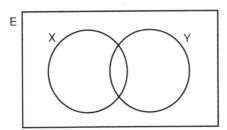

This Venn diagram could illustrate sets such as:
- where E = {the numbers one to ten}
  X = {numbers less than 6}
  Y = {odd numbers}
- where E = {the letters in the word 'days'}
  X = {letters before j in the alphabet}
  Y = {vowels}.

What is the historical background of pictorial representation?

Venn diagrams are named after John Venn who was a Cambridge logician. He used the diagrams in the late nineteenth century to demonstrate the validity of an argument, for example: all B Ed students study mathematics; some females are B Ed students; so some females study mathematics.

Look at the Venn diagram below where S = {B Ed students}, M = {those students who study mathematics} and F = {females} to confirm the third statement.

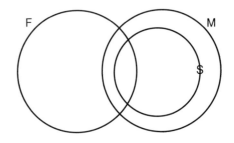

Carroll diagrams are named after Lewis Carroll who invented them. He also wrote *Alice in Wonderland*.

What is the value of teaching pictorial representation?

Presenting data diagrammatically is often a help to the understanding or analysis of it and for explaining it to others. Pictorial data diagrams can be used for finding out about numbers, shapes, words, people, places, plants, animals and all sorts of other things.

What are possible key steps in development for the learner?

1 Sorting into sets

An initial step is simply to ask children to sort items into two sets where one set has a particular attribute and the other does not, for example, sorting cubes into those which are 'red' and those which are 'not red'. The display should focus on separation, for example, in this box (plate, ring) all the items are . . . , while in this box all the items are not . . .

Gradually a system of: collect data, organise it, display it, interpret it, is established as the children decide for themselves:

- what items they will sort and for what attribute
- how they will carry out the sorting
- how they will lay out and label the sets
- what they can find out by looking at the display.

2 Tree diagrams

Tree diagrams are drawn on large sheets of paper and card labels placed on each path. Objects are moved individually along the appropriate paths. The simplest should have a start, two paths (one label for 'having the attribute', the other for 'not having the attribute') and two boxes, whereas the most complex might have a start, two, four, then eight labelled paths and finally eight boxes.

After some experience the children can:

- select items for sorting
- decide on the complexity of the tree diagram
- devise and produce their own labels for each path
- sort the items individually
- display the items clearly in the boxes at the end of the paths
- discuss and agree about wording (a definition) for the objects which are in each box.

Later, older children might be given problems which they can solve using a tree diagram, for example, 'Sort these flat shapes

into quadrilaterals and not quadrilaterals, and into those with or without a diagonal axis of symmetry. Which 'box' on the tree diagrams holds kites? How do you know these are kites?'

3 A tree diagram with representations of objects

The children could sort drawings or names of objects, rather than the objects themselves, using a tree diagram. The drawings or names could be left in position in the boxes by using some 'plastitac'. This makes an attractive poster.

4 Carroll diagrams

Carroll diagrams can be confusing so they should be built up through a number of steps, with the children having lots of experience in sorting at each step. Steps could include:

- two boxes side by side, each with a label used to display, for example, a sorting of different coloured cubes

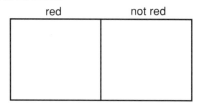

- two boxes, one above the other, each with a label used to display, for example, a sorting of one colour building blocks or beads

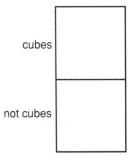

- two boxes, side by side, with labels above and at the beginning of the row, used to display, for example, a sorting of different coloured cubes

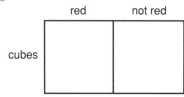

- four boxes arranged in two labelled columns and two labelled rows, used to display, for example, a sorting of differently coloured and shaped building blocks or beads

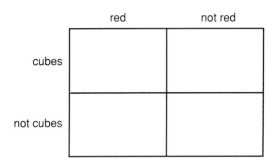

The children can also use representations of objects and then move to devising a Carroll diagram as a written exercise.

The children will become confident in using Carroll diagrams as a means of display if they are challenged to collect, organise, display and interpret their own choice of data, or research questions for this form of display. They need to be aware that a display using a Carroll diagram involves sorting by two attributes, for example, they might want to ask each child in their group 'Do you have a sister?' and 'Do you have a brother?'

Although more complex Carroll diagrams are possible, they are not recommended for primary children. The focus should be on using such diagrams for display of data or to solve problems.

5 Venn diagrams
These are also difficult for some children to understand and are more likely to be understood if they are built up one step at a time. Several examples should be done at each step.
- One set represented in the display by a rectangular sheet of paper and another set represented by a circle of paper. (Later a drawn rectangular outline and circle could be

used.) The shapes to be sorted are placed in the appropriate region.

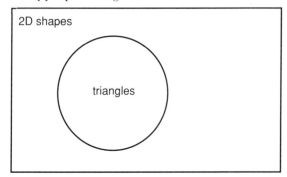

The children should realise that another set is shown in such diagrams – 2D shapes which are not triangles (these are represented by the region inside the rectangle but outside the circle). The children should gradually become aware that the set within the circle needs to contain some or all of the items in the rectangular set.
- One set represented by a rectangular region and two sets represented by circles which do not overlap (intersect). It is easier for the children if they think of the set shown in the rectangle, the universal set, as sorted into two sets where the attributes are of the same type, for example, blue and red, or triangles and quadrilaterals. When interpreting such a diagram, the children will find there is another set, which lies in the region outside the circles, for example, the set which is not labelled but is included in the diagram is '2D shapes which are neither triangles nor quadrilaterals'.

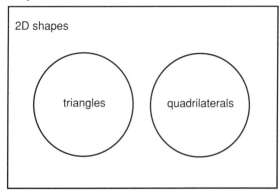

- Some pupils might be able to understand and use a Venn diagram where the sets represented by loops have some members in common and so overlap. Here the pupils should realise that the attributes need to be of different types for example, red shapes and triangular shapes or even numbers and multiples of three.

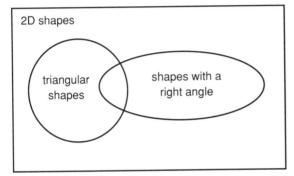

The pupils could learn to use these diagrams for their own data. Some pupils might gain sufficient expertise to select the best diagram for a given situation or problem.

What are appropriate resources for teaching pictorial representation?

Sets of building blocks, 'flat' and 3D shapes, paper and card for pictures and labels, large sheets of paper for floor diagrams, felt pens, boxes, plates, hoops, and string.

What are possible contexts through which pictorial representation might be taught?

Nearly every context would benefit from some information displayed as diagrams, for example, food, holidays, the family, shopping . . .

How might pictorial representation be assessed?

Assess *orally* by asking children to:
- suggest labels for paths/branches of a tree

diagram given a specific set, such as names of some children in the class
- suggest what they might collect data about if it is to be displayed in a Carroll diagram
- interpret what a given region/box in a diagram represents.

Assess *practically* by asking children to:
- sort a given set, for example, books, using a tree diagram with labels such as 'has a picture on each page', 'does not have a picture on each page', 'is about an animal' and 'is not about an animal'
- sort a given set, for example, of words written on cards (book, cook, ball, call, broom, brook, door, dance), to find 'those beginning with b', 'those not beginning with b', 'those with oo' and 'those without oo' and display the results in a Carroll diagram
- place shapes in the correct regions of a labelled Venn diagram.

Assess *in written form* by asking children to:
- draw a Carroll diagram to show the names of the girls and boys in your group
- draw a Carroll diagram to show the names of the children in your group as 'boys' or 'not boys', and 'aged nine' (or another appropriate age) or 'not aged nine'
- sort a given set of 2D shapes to show the set of triangles; draw a Venn diagram to show the set '2D shapes' and the set 'triangles'; state which of these sentences are true
  - All triangles are 2D shapes.
  - All 2D shapes are triangles.

Assess through *problem solving* by asking children to:
- draw a diagram to show the names of British coins and whether they are copper or silver coloured
- draw a Venn diagram to show the sets 'pupils in our school', 'pupils in our class' and 'pupils in our group'.
- draw a Carroll diagram to show the same information as the Venn diagram; then colour regions in both diagrams to show they correspond.

What are common difficulties which children encounter and how might these be overcome?

Direction of reading an arrow – It is confusing for some children to describe a relationship shown by an arrow. Give plenty of practice using objects and arrows with the appropriate words written on them in the correct direction, for example: Children should say: 'The red pencil is longer than the blue pencil.'

is longer than

Interpreting a Carroll diagram – A great deal of practice may be necessary for some children to describe the two attributes of the items in each 'box', that is, assign the labels for the rows and columns correctly. Children may learn to interpret a Carroll diagram and to identify the appropriate 'box' for a given item, but being able to devise such a diagram for themselves is very demanding.

Interpreting a Venn diagram – This is a sophisticated diagram and its use with some children might be best left until they are older. Other children may find interpreting and even devising such diagrams a challenge which they would find interesting.

COORDINATES

What does coordinates mean?

A set of numbers which uniquely identifies a position relative to a set of fixed positions. For example, a street could provide a fixed position with the building number, the floor number and the flat position stated as coordinates such as 46, 2r. Usually coordinates are used on a grid with both a horizontal and a vertical line (axis) providing the fixed reference points.

What are real life examples of coordinates?

Grids, plans and maps are usually provided with fixed points so that positions can be identified by coordinates. The number of digits used can indicate the accuracy of the reading of the position.

What is the key vocabulary for coordinates and what does each word mean?

Cartesian coordinates – a system for labelling positions in space using a pair of axes at right

angles to each other. The axes usually intersect at **the origin** with a value of 0, with positive values being to the right horizontally, and above vertically. To the left and below 0 are negative values. This results in four quadrants being formed with the first being regarded as the one lying between the positive vertical and horizontal values. A position is labelled by the intersection of the perpendiculars from both axes to the position. The first coordinate, known as the **abscissa**, results from the reference to the horizontal axis and the second coordinate, known as the **ordinate**, is derived from the vertical axis.

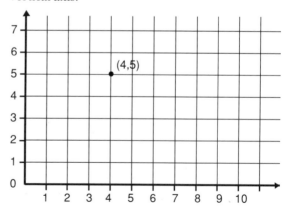

Polar coordinates – where a point is located in a plane by its length from the origin or pole, and the angle from the horizontal axis. Such labelling is not normally used by primary children.

What is the historical background of coordinates?

René Descartes, the French scientist and mathematician, introduced Cartesian coordinates in the early seventeenth century. He wished to systemise knowledge that was clearly self-evident. In Cartesian geometry every quantity has a direction or position attached to it, for example, a line is a number of units of length drawn in a specific direction with reference to other lines. Directions and positions were indicated by using + and −. Descartes did not use the term coordinates. Their invention was by the mathematician Leibniz some fifty years later.

Navigators used the notion of coordinates for position fixing. They created a map on a plane, as if the world was flat, and inserted lines of longitude running from north to south. The line marked 0° runs through Greenwich in England, and the other lines are labelled as west or east of this. There are 360° of longitude. Similarly, the equator is labelled as 0° and there are 360° of latitude which are labelled as north or south of the equator. Positions could be labelled as, for example, longitude 30° E and latitude 40° N.

What is the value of teaching coordinates?

They are used widely in map reading.

What are possible key steps in development for the learner?

1 Knowledge of axes
The children learn about an axis initially as a 'base' line on which to build rows or columns for a pictograph and a block graph. The use of a second axis is developed when bar charts are being drawn. Children should regard axes as fixed reference lines and points.

2 Labelling grid squares
A method of giving each square in a grid an 'address' is to draw a horizontal and a vertical axis. Each square edge length is regarded as an 'interval' on each scale. From the intersection point of the axes, the horizontal scale can be labelled with a reference number – and the vertical scale labelled with a letter. Each square can be considered as the intersection of two perpendicular lines, one from each axis, and referred to by the appropriate letter and number. Some maps use this type of coordinates.

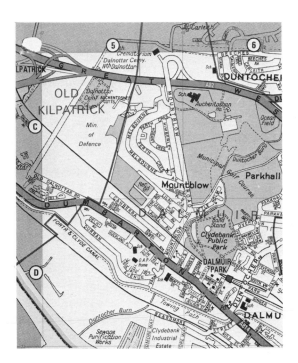

Children can play games where they refer to squares by their coordinates. One of these is 'Battleships' where two players have identical grids. The grids are used on either side of a barrier so that each player can only see his or her own grid. One player positions rectangular shapes which represent a fleet of ships on his or her grid. The other player names squares, using coordinates, pretending that bombs are being dropped on those squares. The other player must admit if a hit has been made. The aim is to sink your opponent's fleet with as few bombs as possible.

Of course, both axes may be labelled using numbers. Such number coordinates can be regarded as an 'ordered pair' as the horizontal number is always given first.

3 Greater accuracy

If a reference for a point rather than a square is required then the grid lines are labelled rather than the squares.

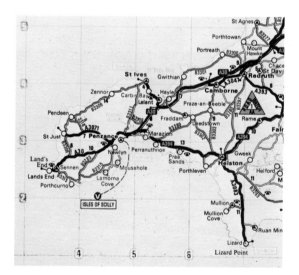

By creating mid-point values, even greater accuracy can be achieved. The children can investigate the coordinates for positions on a range of different maps.

4 Quadrants

Most primary children will focus on positive values for both the horizontal and vertical axes and so use what is only one quarter or quadrant of the coordinate values possible when two axes intersect at right angles. Most should realise that four quadrants exist and be aware of the possible labelling with negative numbers.

What are appropriate resources for teaching coordinates?

Squared paper and a variety of maps. Games like Battleships – either commercial or home-made versions could be used.

What are possible contexts through which coordinates might be taught?

Mapping and orienteering are two appropriate contexts. Computer programs can be used for simulations as well as real maps.

How might coordinates be assessed?

Assess *orally* by asking children to:
- find a pre-determined square where each guess of coordinates is responded to by phrases such as 'the first coordinate needs to be a smaller number'.

Assess *practically* by asking children to:
- walk to a position for given coordinates on a grid marked on the classroom floor or in the playground.

Assess *in written form* by asking children to:
- create a labelled grid which can be superimposed on a given map
- identify positions on a map from given coordinates
- list coordinates for given positions.

Assess through *problem solving* by asking children to:
- draw their own map of the school which includes their own coordinate reference system

What are common difficulties which children encounter and how might these be overcome?

Order of the coordinates – A main difficulty is remembering the order in which coordinates are given. Any confusion can usually be resolved through practice.

GRAPHS

What does graph mean?

A graph is a diagram showing the relationship between sets of quantities or numbers. Coordinates which are used to show positions relative to two axes are one form of graph. The points and arcs of a network are another form of graph. In this section the focus is on what might be called 'Frequency Graphs' where the diagram emphasises the organisation of data into sets and the numerical relationships between or among these sets. Coordinates, networks and other forms of pictorial representation of information are found under separate headings.

What are real life examples of graphs?

Data is often presented as a graph to highlight similarities and differences, for example, the cost of an item from one year to the next. One of the difficulties is that data can be presented in different ways and the deviser of a graph is likely to use a format which best illustrates the point he or she wishes to make, which might be a distortion of the situation.

What is the key vocabulary for graphs and what does each word mean?

Frequency – the number of times that something happens usually within a given time span.
Variable – a 'name' given to set of values, for example, times, heights, pence.
Discrete – separate or distinct.
Discrete data – there are no intermediate values among the data recorded.
Continuous – having a value that changes gradually.
Continuous data – there are intermediate values among the data recorded.

Class interval – data can be divided into groups or class intervals; a class is another name for a set; when each individual item of data cannot be recorded, data can be organised into sets. Class intervals are used on the axis of some graphs.
Pictograph – pictures of uniform size are used to represent data and may be arranged in labelled rows and columns with a common 'start line'.
Block graph – one square represents one unit of data and the units are set out in rows and columns.
Bar graph or bar chart – bars or rectangles of uniform width are used to illustrate data. The length of each bar shows the frequency. The bars can be arranged vertically or horizontally.
Histogram – a bar chart where the width of the bars change according to the class intervals; the area of each bar is a feature. This type of graph is not usually used with primary children.
Spike graph – a bar chart where the bars are replaced by lines emphasising that it is the lengths which are compared.
Point graph – where the line of a spike graph is replaced by the point of highest frequency.
Pie chart – data is represented by slices or sectors of a circle. The size of each sector is related to the frequency it represents. It is particularly useful to compare data represented by a sector with the data represented by the whole circle.
Scattergram – each of the two axes represents a different variable, for example children's ages and pocket money, and points represent paired values; emphasises connections between things.
Line of best fit – paired values are recorded as in a scattergram and then a line is drawn where there seems to be a relationship between the sets of data.

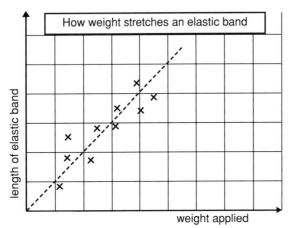

Graph of a function – a line or a curve which shows all the points which satisfy the relationship between two sets of data.

What is the historical background of graphs?

Graphs and maps are similar in that the position of points and the direction of lines are important. The study of astronomy and the realities of war encouraged the development of diagrams where qualities like time and distance were related. Archers wanted their arrows to land where they intended. The path of the arrow, and later of the cannon-ball, gave a new importance to a diagram which showed the vertical and horizontal position at each moment in its flight.

What is the value of teaching graphs?

Graphs are useful to illustrate data so that relationships can be investigated. They are useful to aid word descriptions and as a tool for problem solving. Pupils would find it useful to be able to interpret information presented as a frequency table or graph.

What are possible key steps in development for the learner?

1 A framework for graph work
Regardless of the type of graph, children should be involved in four aspects of graph work:

- the selection and collection of data
- the organisation of the data
- the display of the data
- the interpretation of displayed data.

The children should be involved in a mixture of being shown and finding their own methods.

- *Selection and collection* – sometimes the children should choose what they want to collect data about. They can form the question they will ask and decide who should be asked. On other occasions some or all of these tasks are determined for them.
- *Organisation* – the children could be shown how to record tally marks and produce different forms of frequency tables. They could consider what replies they are likely to get and plan how they will record them.
- *Display* – the children could discuss the different types of graph. They should realise the type of data which can be presented by each type. They could also identify the advantages of using each type. Such experience will help children to select the type of graph which they might use for their own data.
- *Interpretation* – the focus should be on the shape of the graph and then individual features noted. The children should be able to interpret graphs drawn by classmates, the teacher, and other sources. Sometimes they may be led through the interpretation by answering given questions. The children could lead others through the main features of their graph by posing questions. With experience, children may learn to identify the main features in a given graph and describe these without the structure of questions.

Naturally, the children are not always faced with all four aspects for all tasks, and each aspect may be dealt with separately.

2 Three-dimensional representation of data
The data in this instance concerns objects or the 3D representation of objects, for example

favourite toys. Toy cars can be used to represent car makes or colours, paper cups of drink (or coloured water) can be used to represent different drinks. The display is made using the objects or their representations. These may be grouped into sets or may be arranged as columns or rows. The diagram must allow comparisons to be made visually.

If the more formal arrangement of columns or rows is used then the essential learning points in the display are that each object must be positioned so that it takes up about the same amount of surface, and that the columns or rows should have a common starting level. The display might be set out on a grid to help emphasise these points. Labels should be written to indicate a title to explain what the graph is about, and for each set shown.

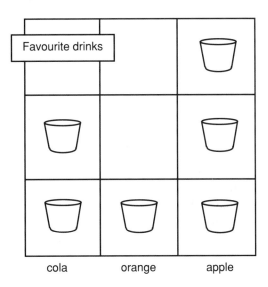

Favourite drinks

cola   orange   apple

3 A pictograph
The data is represented by pictures instead of 3D objects for a pictograph. The knowledge that each picture should take up the same amount of surface could lead to identical pieces of paper being used. Each child, for example, could draw a picture of his or her head to fill a given piece of paper with a feature like eye colour emphasised. The pieces of paper are

then organised into a display with an axis and labels.

With older pupils, this form of graph can be used where each picture or symbol represents a number of objects, for example, one drawing of a human being could represent 100 people, one car could represent 1000 cars, and one house could represent 20 houses.

4 A block graph
The emphasis might be on the children thinking about what responses they will be given as they collect data and how they can organise this to create a frequency table, for example, by recording a tick for each piece of information. The children may find that as well as specified sets, there is the need for a set labelled 'others'. Each unit of data in a block graph can be represented in display by an abstract shape, usually a square.

5 A bar graph
Instead of each unit being displayed by an individual shape, the whole set can be shown by a bar. This may be illustrated to the children by using Cuisenaire unit rods and then exchanging these for the matching rod for the total number in the set, for example, a yellow rod would replace five white unit rods.

When such graphs are drawn using squared paper, the children will soon realise that they need to count the squares to find the value of the bar. If unlined paper were used this would not be possible, so a scale is required for the frequency (how many) axis to show the interval (length) which will represent each unit. As usual, these bar graphs require a title and each axis to be labelled.

Scales can be used where each unit is not recorded but marks might indicate intervals of 2, 4, 6 . . . ; 5, 10, 15 . . . etc. Squared paper is normally used to give easily marked intervals. When a frequency table is being prepared at this stage in the development, the children could be shown how to record five as four vertical strokes with one diagonal across them to make the total easier to count.

Drink	Frequency	Total
cola	~~IIII~~ ~~IIII~~ II	12
orange	~~IIII~~ I	6
apple	III	3

6 Spike graphs

The change to lines instead of bars for each set of data is not a difficult idea, so the focus at this stage could be on the children taking greater responsibility for all four aspects of carrying out a survey. Greater emphasis might be placed on the labelling of the displayed graph. The children could also be given graphs of known types which include errors of presentation so that they can be aware of possible faults.

7 Point graphs

The change to recording only a point, thought of as the tip of the 'spike', instead of a line should be readily understood. This procedure gives the opportunity for the children to realise that points are positioned according to their relationship to each of the two axes, for example, a dot might show William's height is 134 centimetres.

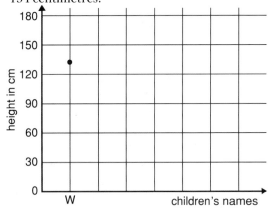

At this stage children could be introduced to data being grouped into 'class intervals'. If, for example, the children wish to include the heights of all the pupils in the class, it might not be possible to identify each pupil separately. The notion could be introduced of taking all the pupils who have a height of, for example, 125 cm to 128 cm, then 129 cm to 132 cm and so on. Such examples link to the finding of the average (mode and mean especially). The use of two axes in these examples could lead to work on coordinates.

8 Pie chart

The children could be shown some examples of pie charts so that they may deduce that the size of the sectors are used like the bars in a bar chart. They can also be directed to realise that too many 'slices' make a confusing diagram so a maximum of about six slices is usually best for this format. The important learning point is to realise that each slice is considered as a part or fraction of the whole circle. For example, if the whole circle represents Mark's pocket money for a week, a sector would indicate what fraction of it he spends on a game for his computer. Initially the children could use circles marked out in fractional parts with appropriate data. Later the children could use circles marked with 10° or 20° degree angles. A few children might use a circular protractor to create their own chart.

The children should discuss labelling as the clarity of a pie chart can be spoiled by the amount of written words. Usually numbers indicating frequency should not be used on the chart as the size of the sector illustrates this.

9 Scattergram

Here the two axes usually show scales for continuous data. A useful previous experience is work on coordinates where both axes are number lines. This type of graph is useful to see links between two sets of data, for example, age and height, age and amount of pocket money.

Some simple ideas about sampling might be discussed when the children are considering what data to collect and how meaningful it is. They could become aware of a sample as a selection of possible data.

10 The line of best fit

This graph is suited to experimental data, for example, if the children are finding out how the length of an elastic band is related to the weight suspended from it, they are likely to obtain points which line near a straight line rather than on it. It is quite sophisticated for primary pupils to consider where a possible line might be drawn to relate the measures.

They are likely to need a great deal of practice to use this type of graph for themselves and this would overemphasise this form of graph. However, it is useful for them to record results as a scattergram and the possible 'line' be shown to them and discussed.

11 Graphs of functions

The children could be given straight line and curved graphs to interpret. In this instance the focus should be on what the shape of the graph means, for example, a steeper slope to a straight line using the same axis means a greater increase in the vertical measure than the horizontal. For example, a graph of the 6 times table has a steeper slope than that of the 3 times tables (when the horizontal axis shows the number being multiplied and the vertical axis shows the product) because the difference between the products is greater.

Graphs of stones falling, balloons rising and balls rising and falling where time and vertical height are used on the axes, produce curves which are interesting to discuss.

What are appropriate resources for teaching graphs?

Three-dimensional objects, toy or model objects, plain and squared paper, coloured pencils/pens.

What are possible contexts through which graphs might be taught?

Data can be collected in most extended contexts to inform the children about a social, fictional or factual situation. The children particularly like to find out data about themselves, their families, their town and their country.

How might graphs be assessed?

Assess *orally* by asking children, for a given graph, to:
- identify what is missing
- identify mistakes
- name an appropriate title
- name an appropriate label for an axis
- describe the important features.

Assess *practically* by asking children to:
- set out a 3D graph to show whether children in the group wear more digital or analogue watches
- direct classmates to produce drawings for a pictograph of their favourite activity outside school
- carry out a survey and produce a frequency table about the number of hours of TV each classmate watches in one week
- set up an experiment to find the length of the longest jump which can be made by someone in their group; record the results as a bar graph for a wall display
- set up an experiment to show the distance a classmate can move a large box which has been filled with different weights (about 20 to 60 kg) using his/her feet; show the results as a graph.

Assess *in written form* by asking children to:
- draw a specified graph to display given data.

Assess through *problem solving* by asking children to:
- carry out a survey of their own choice to find out information about classmates.
- sketch the shape of a graph for the journey to buy milk.

What are common difficulties which children encounter and how might these be overcome?

Labelling – Some children are likely to forget to label parts of a graph. Experience at trying to interpret graphs which are poorly labelled should highlight the importance of this feature.

Interpreting – Practice at interpreting line and curve graphs can lead children to understand how the relationship between variables alters the steepness of the slope etc.

DATABASES AND SPREADSHEETS

What do database and spreadsheet mean?

A database is simply a collection of data. This could be in a book, on cards, or on a computer disc. Data can be updated by entries being crossed out or added in. When a book is used, updating is likely to produce rather poorly presented records which may be misinterpreted. If cards are used, the data might be updated by replacing individual cards which is effective for good records, but laborious. Data stored on the computer disc is easily updated. However, the main strength of using a computer is that the computer can search, sort, display and sample the data at the press of a key.

A spreadsheet is a special type of database where information is set out in a table with labelled columns and rows. It is just like the tables you see in holiday brochures where the columns may be labelled with the number of nights and the time of year, and the rows labelled with individual hotel names. In the 'cells' of such a table you find the various costs. The great advantage of using a spreadsheet is that if numerical changes are required, for example, if the holiday company can offer a reduction on all holidays in the table of 10%, then the computer can recalculate all the prices at the press of a key.

Both databases and spreadsheets usually include the facility to draw graphs in a variety of formats for selected data.

What are real life examples of databases and spreadsheets?

Spreadsheets are used in business to calculate wages, to keep a control of stock etc. Databases are used by libraries and museums to keep collections up to date. It is also quite usual for people to keep records of members of a club and items in a collection on a computer disc.

What is the key vocabulary for databases and spreadsheets and what does each word mean?

Record – each piece of data.
Field – the heading under which the records are grouped.
Menu – a list of commands which can be requested or a list of the files available.
File – a collection of records.
Search – a command used to select records to fulfil particular conditions.
Sort – to change the order of records in some way.
Display – to show the required records.
Graphics menu – will offer a range of different graphs in which the selected data can be displayed.

Statistics menu – will offer to carry out a calculation to find the mean, median, standard deviation and other statistical calculations.
Sample – if a data file is very large it may be desirable to use a smaller number of records. These can be selected in different ways, for example, randomly or according to specified conditions.

What is the historical background of databases and spreadsheets?

In the seventeenth century Blaise Pascal, a Frenchman, invented the first mechanical adding machine. In the nineteenth century, Charles Babbage an Englishman, created a 'difference engine' which was an automatic calculator. He later went on to devise an 'analytical machine' which was to have the capacity to read data from punched cards, to store this and print it out. However, the 'analytical machine' was not completed. In the United States, Herman Hollerith was convinced that a machine which could read information on punched cards would be the answer for the census data in 1890. Some years later he did invent an automatic data processing machine. He also began a company which was later to become the International Business Machine Corporation (IBM) to produce these machines. John Von Neumann completed the first computer which was capable of storing its programs internally and working at electronic speed in 1952. Rapid changes are still taking place in the development of computers.

What is the value of teaching databases and spreadsheets?

Since computers have become transistorised, they have become smaller and cheaper. As more are sold they become even cheaper. They are used widely in government, business, and shops and are becoming popular in the home. The credit card systems need the back-up of effective computerised records. There seems little doubt that gradually most data storage will be computerised.

What are possible key steps in development for the learner?

1 Using a database

There are many databases specifically created for school children to use. These databases are often supplied with prepared files based on geography place names, historical events or animal descriptions. The children could learn how to use the base to answer questions and solve problems. The screen menus in many of the programs will lead the children through steps to use the program successfully, but the capability of a program should be discussed with the children and they can develop 'rules' about how to carry out procedures. Children find out the meaning of phrases such as 'is the same as', 'comes before' and 'contains' by carrying out searches. Some data may be prepared specifically to fit with a theme the children are studying to allow them to have such practice.

2 Creating a database

The children may require help in setting up the fields for a database and some instructions may be prepared to lead them through the initial stages. The instruction manual should be helpful for teachers doing this. Many children will simply find out for themselves what works and what doesn't, although it can be very frustrating for them if they have entered a large number of records and then discover that these are not in the correct format or can't be 'saved' on disc.

Some children will see the database program as an opportunity to keep records of themselves, their library books, events in the school session, the tuck shop stock, the variety of food in the school canteen and its popularity and so on.

3 Using a spreadsheet

Pupils can be introduced to spreadsheets

particularly for interpreting numerical data. A spreadsheet program could be particularly useful when the pupils are working with percentages. There may also be suitable applications arising from theme work like holidays.

What are appropriate resources for teaching databases and spreadsheets?

Software packages which are 'user friendly' for the children.

What are possible contexts through which databases and spreadsheets might be taught?

Practically any subject background has data which can be collected, stored, sorted and displayed.

How might databases and spreadsheets be assessed?

Assess through *problem solving* by asking children to:

- devise a database for the main rivers in Scotland or England
- devise a database for the main characters in the novel you are reading together.

What are common difficulties which children encounter and how might these be overcome?

Entering data – Children should work on this together as it can be very laborious and boring. The data will also require to be edited as spelling or typing mistakes may lead to errors. If the children work with one or two classmates they should be able to edit each other's work and prevent some mistakes.

Creating fields – The children should be encouraged to think out their fields before using the computer. Discussion can result in a more effective list of headings and some insight into the likely records and sortings.

SECTION TWO
TEACHING AND LEARNING

S E V E N

Introduction

In this section possible approaches to learning and teaching mathematics are considered. The ideas suggested sometimes overlap and the list is not a finite one, but it will be useful for teachers to review what approaches might be used. Many teachers may be using all of the approaches already in other subjects, but may not have thought of using them for mathematics.

As in Section One, for each of these approaches, a number of questions are posed and answered. Here is a summary of what you might find in the response to each question.

What do we mean by . . . ?

An explanation is given of what is meant by the title given to the approach.

What organisation is involved?

Here the response is given under the headings of Organisation, Location, Communication and Feedback.

Organisation – indicates whether the children should work individually, as a group or as a class. There are suggestions about how groups might be formed. Sometimes there is advice on what resources are required and how these

might be used. Thoughts about timetabling might also be included.

Location – suggests any special problems which might arise in helping to ensure that children work effectively with maximum freedom, but with minimum disturbance to others.

Communication – focuses on instructions which you may need to give the children and/or their expected behaviour.

Feedback – suggests follow-up material from you and/or from the children themselves.

What does the teacher hope to achieve?

The aims which you could have for using a particular approach are listed.

What does the teacher do?

This section is subdivided into: Initial teaching, Encouraging the approach as part of learning, Setting tasks for the children, Coping with difficulties, and Record keeping.

What do the children do?

Some objectives which might be achieved by children through the use of the learning approach are given.

EIGHT

Focus on children listening and talking

WATCHING AND LISTENING TO THE TEACHER

What do we mean by 'watching and listening to the teacher'?

All children spend time each day watching and listening to the teacher. This is an important part of teaching and learning. For much of the time the children do this automatically, often not really giving their full attention. If you want to focus their attention, some device is usually necessary, such as gathering a group of children at a maths table, or producing a visual aid which will capture the attention of the group or the class. For example, if children are to learn a series of steps to use a balance effectively, a demonstration at the maths table could show the correct order of actions and prevent children adopting habits which they might have to unlearn later.

Another group situation might be to use Base Ten pieces systematically in a subtraction calculation, for example, by using a notation card where the pieces which represent the larger number are exchanged and repositioned to show the number 'taken away' and those 'left'. Any exchanged pieces might be placed in a box. (See diagram opposite.)

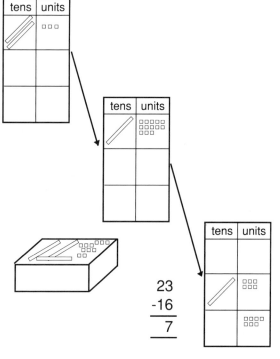

Listening skills in mathematics might be developed by playing a guessing game with the class where clues are given for the children to identify a number or a shape. Children could be asked to put up a hand for the correct name, correct pronunciation, the correct numerical answer. For example: Is this shape a rhombus, a rhomnibus or a remus?

Stories which feature mathematics in a context could be told to the group or class with a follow-up task, such as asking them to:
- recall a number, a direction or a shape
- collect clues about a number/direction/shape
- follow directions on a map
- draw a sketch of an object/person/place.

What organisation is involved?

1 Organisation

For the 'watching and listening' approach, the teacher is likely to demonstrate or talk to all the children in the class or one particular group. While there is likely to be some interaction with the children, this approach is teacher dominated. Naturally if the children are to watch, they should be gathered together in such a way so that each can see clearly, and if the children are to listen, other noises should be minimised.

2 Location

This will depend on the content which may demand working beside the sink, on the floor or out of doors. It is important that the children should be comfortable and distractions avoided.

3 Communication

The children could be told, for example, 'I want you to watch very carefully because the order of the steps when you set up the balance is important.' This focuses their attention not only on the content of the lesson but also on their expected behaviour.

4 Feedback

The children could be made aware if they have watched or listened effectively by monitoring performance in a follow-up task. Praise should be given where they show they are implementing the advice given.

What does the teacher hope to achieve?

You would expect the children to adopt specific ways of working where it was believed to be beneficial to the child to do so. Through planned listening you hope that children will learn to be selective in the level at which they listen. In the classroom and at home, children discount a level of background noise, conversation which is not directed at them and noises in the street outside. Many children find it necessary to learn to listen and gradually increase their concentration span for absorbing information.

What does the teacher do?

1 Initial teaching

Most teachers of young children adopt an approach which settles the children, for example, 'Are you all ready to listen? Good, then I'll begin.' With older children, you could label activities on the group's programme of work as, for example, 'doing', 'talking', 'reading', 'writing', 'watching' and 'listening'. This identifies the expected behaviour for the work involved. There should, of course, be 'time-out' spells when the children can relax and no specific behaviour is expected.

2 Encouraging 'watching and listening to the teacher' as part of the teaching/learning procedure

You need to be realistic about the time children can concentrate on watching and/or listening. This depends on the age and ability of the children as well as the practice they have had in doing these activities. Your performance should be well planned and motivating for the pupils, perhaps because of the context, and carried out with interest and/or enthusiasm. Use of a visual aid, for example, a poster, a model, one of the children, an extract of video, or an oral aid such as a taped speaker(s), an extract from a radio broadcast, or taped sounds, can make a good beginning to catch the children's attention.

3 Setting tasks for the children

Here are a few examples:

For watching

- *Sorting 2D shapes*
  The children are asked to guess why some shapes are put into the hoop and others are not as you sort 2D 'flat' shapes one at a time,

into sets. You could sort a set of shapes into those with more than three edges and a set of shapes which do not have more than three edges.

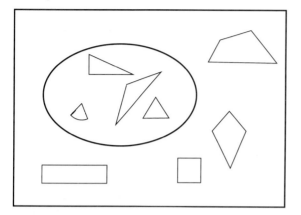

- *Sorting numbers*
  The children watch as you move the numbers 10 to 20, each written on a separate card, along a tree diagram. Their task is to decide what each path means.
- *Subtraction with Base Ten pieces*
  The children have already carried out subtraction using the decomposition method with Base Ten pieces. They watch as you make links to each step in a written calculation.

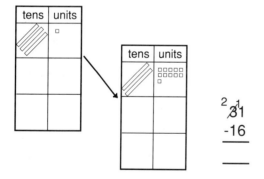

For listening

- *Missing words in a story*
  The children each have a set of flash cards showing the words 'tall', 'high', 'long'. You tell a story which has missing words. Each time 'gubble' is said the children know there is a

missing word and that they should choose the most appropriate flashcard and hold it for you to see.

- *Guessing the 3D shape*
  You describe a 3D shape and ask each child to put up a hand when he or she is able to guess the name of the shape.
- *Mistakes*
  You describe a shape and the children are asked to listen for mistakes made in the description.
- *Movement of shapes*
  Each child has a scalene obtuse-angled triangle. You will give instructions for movements, for example, a quarter turn clockwise rotation, then a translation, then a reflection and so on. Each child should move the triangle, if possible out of the view of classmates (for example, within a box lid). It will be interesting for each child to compare the final orientation of his/her triangle with those of classmates.

4 Coping with difficulties which might arise
The main difficulty is likely to be that some children find it hard to concentrate for any length of time. They may need to be involved initially in short watching/listening sessions with these being gradually lengthened. You must also be sure that what the children were watching or listening to was set in an interesting context or was perceived by the children as having a purpose.

5 Record keeping
It is useful to note children who lack concentration. This difficulty should be discussed with the child and the parent so that the child tries to improve and is rewarded when he or she does so.

What do the children do?

The children should realise that there are times when watching and/or listening to the teacher is important.

TALKING WITH THE TEACHER

What do we mean by 'talking with the teacher'?

Each child should have the opportunity to talk to the teacher, both as a member of a group and as an individual. The child should be aware that you ask questions to find out what has been understood, and what has not, so that help can be offered where it is required. The child should also, of course, have the opportunity to ask you questions.

Talking with the teacher can happen spontaneously. However, it is also useful to plan such sessions.

What organisation is involved?

1 Organisation

Finding time to talk to individual children is the most difficult part of any teacher's job. Fortunately through teaching a group, you have the opportunity to interact with individual children and the time to identify needs. To meet these needs is more difficult as it may require talking with the individual child rather than talking with the group.

You can monitor how time is being spent so as to use it as profitably as possible. For example, Do one or two children monopolise too much of the time? Are there some children whose work has not been discussed for a week? Or has a child asked a question which you said you would discuss later and later did not materialise?

Planning the time does not solve all the difficulties but it helps. Some teachers find it useful to post a programme of work for each group, labelling the activity at which you will be with them, for example, using the letter 'T'. There could also be a time allocated to individuals where the children can come, or 'book' time, to talk individually with you about

work, or you can approach individual children. This sounds rather formal but in operation it should not be. The reason for planning time is to make sure that such a time is considered important and not forgotten.

2 Location

Many teachers move to the group so that the children work at their tables/desks with their textbooks and equipment on hand. Other teachers ask the children in a mathematics group to come to work at a table within a 'maths corner' and use the resources which are kept there. Some teachers ask the children in a group to gather at the blackboard/whiteboard, large book, sink, weight table etc. When working with an individual, you may go to where the child is sitting or ask the child to come to some quiet corner.

No matter where the location, whether working with a group or an individual, it is advisable if you place your chair with its back towards the wall of the classroom so that you can have the rest of the class in view.

3 Communication

You are likely to use a different voice when speaking to the class, to a group and to an individual. This difference cues the children into the situation and the type of response expected.

4 Feedback

Many children are uncertain about the role of the teacher. They should realise that you want each of them to learn and will do everything possible to make that happen. It should become obvious to the children that their effort to do well, regardless of how successfully, is met with encouragement and that difficulties are listened to with sympathy. Praise from the teacher is essential. Each child should know that he or she is important to you.

What does the teacher hope to achieve?

You will realise that talking with the children is the best way to find out their understanding and their difficulties. Fortunately, for much of the time, the children have a shared need, so class and group discussions are appropriate. However, when a child makes a mistake, finds something difficult, or does not seem to understand, a chat with the individual is the beginning to finding a solution.

You should try to know each individual as comprehensively as possible, and it is important to try to meet the child's family whenever possible, so that they can be involved in helping the child.

What does the teacher do?

1 Initial teaching

In research work, teachers sometimes are required to focus on the work of one or a few children who are experiencing difficulties. It is interesting that with the extra attention many of these children make excellent progress. Your main task therefore is to make sure that each of the children in the class experience this feeling of 'extra attention' at least some of the time.

Teaching should be aimed at the individuals even though the organisation involves a group or the class. Find out by asking a few questions during, and at the end of a lesson, if some children seem to have understood. Check on the understanding of others through a follow-up task. Keep a record of an individual's short-term learning. Note only the significant, that is, the children who don't know or understand, and what the difficulty is. More formal assessment could act as a check on long-term learning.

Talking with the teacher could be described as 'teacher orientated' as the initiative lies with the teacher to encourage and allow the pupils to talk.

2 Encouraging 'talking with the teacher' as part of the teaching/learning procedure

The children should be aware that there are times for social chatter with the teacher, usually as the class gathers in the morning and as they get ready for break, for lunch and for home. The children should also realise that if they don't understand in a lesson there is time to catch your attention and ask a question. However, the children should equally be aware that there are times when you should not be disturbed.

When you are ready to check on work being carried out by others, an important decision may be whether to depart from teaching plans to cope with a child's difficulty immediately, or to explain to the child that a discussion about his or her difficulty will take place in the afternoon/tomorrow or at some time when the individual session can be integrated into the class management. With young children it is usually best to find time immediately as the difficulty is likely to be forgotten. This simply indicates that planning must remain flexible to encourage the children to express their difficulties.

3 Setting tasks for the children

It is useful to have sessions where you ask the children in a group to tell you what they find difficult, but much of this aspect of 'talking with the teacher' should be child-initiated. You need to ensure that time is available for this and when a child takes the opportunity, you listen and then help.

You should also plan interaction with children so that they can learn to express thoughts, listen to others, learn to modify their ideas and to reach agreement. Here are some suggestions for mathematical discussions:

- *What is volume?*
  You could begin the discussion by asking the question 'What is volume?' The children should be encouraged to explain volume in their own words, agree and disagree with each other until some definition can be

arrived at. The appropriateness of the definition can be tested as you challenge the children's thinking, for example, by asking 'Do you have a volume?' In such discussions you should intervene only when necessary to encourage individuals to participate and/or to direct the argument along a more profitable path.

- *Introducing a new game*
  A game may be introduced with a rule card. However, you may wish to be present to establish good habits of play, for example, that the dice is collected after a throw, returned to the shaker and handed to the next player, or that a number along a track is counted by beginning in the 'next box'.

4 Coping with difficulties which might arise
Many children find it hard to express their difficulties and needs. Most just require more opportunity to talk to the teacher. A few need a great deal of encouragement and understanding to communicate clearly.

5 Record keeping
You could note difficulties which children identify for themselves as well as those you have found. It is rewarding to find evidence in the future that the difficulty has been overcome and the child can be given confidence with comments such as 'Good, you don't find that difficult any more.'

What do the children do?

The children should thrive in an environment where there is understanding, encouragement and the challenge to do well.

TALKING WITH CLASSMATES

What do we mean by 'talking with classmates'?

The children work in a group, independently of the teacher, with a specific remit which requires them to talk together to:
- interpret what they are asked to do
- decide on how to proceed
- agree on a response
- communicate the response.

What organisation is involved?

Organisation
The children could be organised in groups of three or four so that there is scope for everyone to be involved. For example, one child could be allocated, or choose from a limited selection, a discussion task and then ask two or three classmates to work with him or her. In this way an organisation is set up where children have a sense of responsibility – they choose to work with each other and on a specific task. It is probably best to have only one group, possibly sub-divided into two or three smaller groups, working in this way at any one time, otherwise you will find it difficult to monitor what the children are doing and will not have enough time to follow up with questioning or a report.

Location
As the group is working independently, the children can be located where you can best monitor their progress, yet allow them the freedom to talk naturally. In some classrooms this is possible in a library corner, in others, a table and chairs to accommodate the children can be placed either in a walk-in cupboard or just outside the classroom door. Some schools have open areas, a school library and/or a spare room, which can accommodate working groups yet still allow some measure of supervision.

Many children, of course, can accept the responsibility to work without direct supervision.

Communication

The task should be structured so that the nature of the response is either stated or the children know they have to decide upon this. The smaller and more competent the group, the less structured the task needs to be and vice versa. Responses might be:

- a list of oral or written decisions
- an informal report to the teacher
- a presentation to classmates.

Feedback

The children should be told if their discussion skills meet with set objectives and, if not, how they can improve. This means you need to find time to listen to a sample of the discussion as it takes place, or through the use of a tape-recorder.

What does the teacher hope to achieve?

The emphasis here is on children talking about mathematics. So often children appear to understand mathematics vocabulary when the teacher uses it, but their own oral mathematics vocabulary is extremely limited. This often leads to difficulties in developing a mathematics reading vocabulary. Children also have the opportunity to learn through applying their knowledge of mathematics. This should consolidate and extend their understanding as they think through aspects of mathematics and express them to others.

What does the teacher do?

Initial teaching

This could mainly be carried out with a group, possibly as a 'talking with the teacher' session.

You could identify a list of behaviours which the children are required to adopt and develop, for example:

- being able to take an active part in a discussion
- selecting appropriate knowledge
- expressing this, using materials and diagrams where these are helpful
- listening to what others say
- encouraging others by asking what they think
- being able to report on what others say
- being able to comment on what others say in a constructive manner
- being able to interact with others by adapting the idea of a classmate
- realising when to press their own viewpoints and when to give way to another opinion or suggestion.

You could focus on each of these behaviours, making children aware of them and giving them a procedure to follow. For example, if you wished the children to gain practice in selecting appropriate knowledge, the individuals in the group could be asked to write about 'What you know about even numbers'. The individual lists could be discussed together so that all the children realise what is relevant and how their combined knowledge is usually much greater than that of just one of them.

If you wish to give the children the opportunity to learn to give a report, individuals in the group could be asked to prepare and tell the others about a specific concept or skill, for example, 'What is a triangle?' or 'How do you find the difference between two numbers?' When the children have listened to a few reports on the same remit, they should be asked to say what was the most interesting thing said and what was the best part of each talk. In this way the children might evaluate the content and the method of communication. When the children are 'practising' each behaviour, that is 'talking with classmates', they work without teacher intervention.

Encouraging 'talking with classmates' as part of the teaching/learning procedure
You might like to show each discussion behaviour as a poster, and plan specific tasks so that each group have the opportunity to practise one behaviour at a time. For example, the behaviour focused upon might be 'being able to comment on what others say in a constructive manner', so the task set could be that one member of the group describes a given 3D shape while the others each write one suggestion to make the description better and then, in turn, explain their suggestions to the speaker. Possibly one group could try this at each mathematics time during a week, and at the end of the week you could ask the children about how they made suggestions and how they reacted to advice from classmates. Children could experience a 'talking maths' task suited to their age and ability at least once each week.

Setting tasks for the children
The tasks are essentially oral although the children may like to write informal notes for themselves or make a more formal record or visual aid for a presentation to the group or class. Example tasks for a group could include:
- *Making a list of shapes*
  Identify all the different mathematical shapes you can see in this picture.
- *Making up a story*
  Make up a mathematical (number, measure or shape) story which one of you can tell to classmates using the character Mr Tall (this car, this puppet, this picture etc.).
- *Devising instructions*
  Agree instructions which each of you will use to tell a classmate how to use a ruler (a pair of compasses, a clinometer, a spring balance etc.).
- *The price of biscuits*
  Decide on a price for the biscuits being baked by the other group (a ticket for a class outing, a ticket for a school concert, items for a jumble sale etc.).

- *Interesting maths*
  Agree about the two most difficult (interesting, enjoyable etc.) things you have had to do in mathematics this term.
- *Guessing game*
  Devise a guessing game where you give your classmates one clue at a time to identify a shape (a number) you have selected.
- *Mental addition*
  Agree instruction steps to tell classmates how to add numbers like 9, 19, 29 . . . mentally (multiply by 10, 20, 50 or 100; divide by 10, 100 . . .).
- *Rules for a game*
  Read the rules and play this new game, then agree how to explain the way to play it to others.

Coping with difficulties which might arise
There are likely to be children who say little or nothing at all, and others who dominate the discussion. It is useful for the quiet child sometimes to be:
- the leader of the group with specific instructions to ask questions such as 'What do you think?', 'Do we all agree with that?'
- in a group where one or more of the others are particularly good at 'encouraging others to talk'
- in a group where each child has a supply of one colour of straw and each oral contribution is recorded by the children putting one of their straws in a communal holder.

The dominant child could be:
- asked to work with other dominant children
- asked to be the leader who encourages others to talk
- required to keep a written record of what the others discuss.

All children would benefit from hearing a sample of one of their discussions where you, and possibly other children, comment on their performance related to a particular behaviour. Children should know what they do well and what they could do better.

Record keeping

A checklist could be kept to show the behaviours in which each child has been monitored and their level of confidence and fluency. Some discussions could be taped using tape labelled with an individual name, so that there is at least one oral mathematics recording for each child each term.

What do the children do?

The children should be aware of the communication skills that they are learning and that these are applicable to any subject. You could ask children to classify oral mathematical remarks to identify if these are descriptions, instructions, explanations, questions, opinions or suggestions. You could also ask them to comment about their own performance in a discussion. They might be guided by questions like these:

- Did I say too much?
- Did I say enough?
- Did I have any good ideas?
- Who else had good ideas?
- Was I good at telling the others about an idea?
- Did I really listen to what the others said?
- Did I say 'That's a good idea' to anyone?

Later pupils could be asked to comment constructively about the performance of others.

DEVISING AND ASKING QUESTIONS

What do we mean by 'devising and asking questions'?

In this type of learning the children are required to think out questions which they ask the teacher or their classmates.

What organisation is involved?

Organisation

It is possible to introduce a procedure for asking and devising questions either through a class or group lesson. When the children realise what they have to do they can work both individually or in groups of three or four. It is recommended that no more than three groups are involved in this procedure at a time, otherwise follow-up can become too difficult.

Location

The children should be able to carry out any group work in the classroom, although a more private working area such as a library corner or in the open area would be ideal if a group is involved.

Communication

Questions could be written by the children as they make them up. They can be presented either orally or written for you or for classmates.

Feedback

The questions should be answered, sometimes by you, but more usually by classmates. This requires you to meet with the questioners, and the children who have attempted answers, to discuss the nature (interest, feasibility, or practicality) of the questions, the clarity with which they are expressed, and any difficulties in making responses.

What does the teacher hope to achieve?

Teachers consolidate and extend their understanding as they think through questions to ask the children. It seems sensible to provide the same learning opportunity to the children. They should find themselves applying their knowledge and extending it.

What does the teacher do?

Initial teaching

You could introduce this form of learning by playing a guessing game. For example, you could tell the pupils 'I am thinking about a shape/number, ask me questions to find out which shape/number it is.' Initially, the questions for this are likely to be restricted to those which are answered by 'Yes' or 'No', but such questions are appropriate to enable the children to gain confidence. Follow this with a child taking the leading role by selecting a shape/number and then answering the questions posed by the others in the group. When you allocate written examples to children or check answers to written work, the grading of the questions could be commented upon. Ask the children which was the easiest and which the most difficult and why they came to those conclusions. Such discussions would help equip children to make up written examples for themselves and for classmates.

In discussion and problem solving, children often think of a question in their minds and begin to answer it without making the question explicit. This possible action should be explained to them so that they focus on devising the question, not answering it. You could also make sure that 'devising the question' is clearly expressed as a task for one group while finding an answer is a task for a different group. The group devising the question could follow a procedure where they write the question and focus on how it is worded and if it is easy to understand. The advantage of a group rather than an individual devising the question is that they need to make the question explicit to discuss it together. However, some individual work is also required so that each child has practice.

Encouraging questions as part of the teaching/ learning procedure

(a) Questions as a basis for discussion

When something new is being introduced to the children, you could develop the habit of telling them what the focus is for the lesson and asking them;

- what they already know
- what they want to know.

You may find that a 'talk' lesson develops into an interesting discussion as the information which the children offer can contribute towards a learning structure where they ask questions to find out more. This might not be as successful as you would like to start with, as the children will need practice in identifying what they want to know and in devising the correct questions to provide them with meaningful answers. This procedure is likely to demonstrate the value of group teaching as the knowledge the children already have and the questions they will ask will make the time spent with each group take a different form.

(b) Questions made up by others

Children could be encouraged to constructively criticise questions in their mathematics books. Questions which are misleading could be identified, discussed and adapted so that their meaning is clearer.

(c) Collecting and handling data

Whenever possible encourage the children to identify what they should collect data about and to devise their own research question.

(d) Reference material

Children could decide for themselves which reference materials might be useful to find answers to their questions. They should learn the value of a mathematics dictionary, of mathematics reference books, and encyclopaedias as part of their mathematics work. They are likely to find that the answer to one question often leads to another question.

Setting tasks for the children

Tasks like those which follow can be set orally or in writing and can be given to a group or an individual.

- Make up word problems for these answers:
    (a) 24 boys   (b) 16 girls.

- Here is a new game. Make up three questions to ask about how to play it.
- Make up question cards for this track game.
- Make up six multiplication calculations for a classmate to do.
- Make up five questions for a classmate where a calendar is used to find the answers.
- Make up six questions where a classmate uses a calculator to find the answers.
- Make up questions to ask about this/your graph.
- Make up questions to use with the database we are going to build up on the computer.

Questions made up by the pupils could be collected into books for classmates and pupils in other classes to use. Answer books could be produced too.

Make resources available
The basic materials are paper/card and pens/pencils for the children to use to draft questions and then sometimes to make a question card for a classmate. Questions may also be devised on the computer screen and the final version printed.

Difficulties which might arise
Confidence – some children will not offer a question and will need to be encouraged to do so.
Ability – some children might need to be given a stem initially, for example, 'Make up a question to complete each of these':
– What is . . .
– What does . . .
– Where is . . .
– Why does . . .
Facility – some children may need time to express their question orally so occasionally they should be taken aside on a one-to-one basis as a follow-up to the group work, so that other children do not become impatient. Some may require help with writing the question and find the word processing package on the computer makes the task more interesting and easier for producing several drafts.

Record keeping
Written questions made up by a group or an individual could be kept as part of a child's folder of work. The questions could have an informal note detailing any strengths or weaknesses about the child's ability to structure questions.

A checklist may be marked to show if the child is able to:
- make up a question beginning with an interrogative pronoun or adverb, for example, 'What is . . . ?' 'Where is . . . ?'
- devise a range of questions appropriate to a given situation
- communicate his/her question effectively
- answer his/her question
- accept alternative appropriate answers to his/her own
- interact with those offering answers to extend the original question.

What do the children do?

1 Following a procedure
The children could learn a procedure, such as:
- think about what question to ask
- decide on a good word to begin the question
- try completing the question in different ways
- use the wording which makes the question clearest
- remember to put the question mark at the end.

2 Asking and answering questions
You may wish to separate the task of asking a question and answering a question by making these tasks for different groups. Each group should be clear what they have to do. However, the children making up the questions should adopt a final stage where they consider what they might expect as an answer to a question. If the question asks 'Can you . . . ?', the children should realise that the only answer they can expect is 'Yes' or 'No'.

They could be asked to produce an 'answer card/book' for their examples. This makes them consider how examples could be labelled as well as how the answers could be expressed.

3 Making up calculations
As well as making up word problems and questions for discussion, the children should be given the task of making up numerical questions. Some children will find it difficult to think of even one or two examples, while others can accept the challenge of:

- involving as many different facts as possible
- of grading the examples from easy to difficult
- of including particular difficulties of place value, having a zero etc.

If the examples are presented as a workcard the children should be encouraged to think about layout. If the children produce answers, they need to consider how questions are numbered to make the answers correspond.

READING

What do we mean by 'reading'?

The term reading is used here to cover a range of skills including:
- saying printed or written words aloud with understanding, fluency and expression
- identifying and researching words of which the pronunciation and/or the meaning is unknown
- looking at the words and interpreting their meaning
- being able to communicate a personal interpretation of an extract
- being able to summarise the contents of an extract.

Such skills are usually taught under the heading of language, but there is a mathematics vocabulary which children might only meet in mathematics texts, so reading skills should be met in a mathematical context too.

What organisation is involved?

Organisation
The skills involved in reading are learned by each child. It is usual to have children of similar ability in reading grouped together, but tasks given to such groups require, in most instances, an individual response. The individual

responses may be given as part of group work as well as when working with the teacher. 'Reading' of mathematics vocabulary can, of course, be carried out as part of language, environmental studies, and project work, as well as during mathematics lessons.

Location
There are no demands on a special location for much of this work. However, if children are reading aloud individually the setting has to allow them to be comfortable, easily heard and yet not liable to cause an interruption to the work of others.

Communication
You can ask children to read aloud to yourself, to a tape-recorder, to an adult helper, to a classmate or to a group of children. Interpretation and summaries of written extracts can be given orally as well as in writing. Unfamiliar or unknown words can be identified orally or in a written list. Pronunciation of such words is, of course, oral work. Establishing the meaning of new words could be part of a reference skills task in which a mathematics dictionary might be used as well as a language one and encyclopaedias. After the children have carried out such research for themselves, new vocabulary could be the focus of a group

discussion so that the children have the opportunity of developing an understanding of each word and of adding it to their oral vocabulary.

Feedback
The children should realise that they are building up a mathematical vocabulary and that some words, for example, 'round', 'kite' and 'volume' which they may use in conversation to convey one meaning, have a different or more specific meaning when they are used mathematically. For example, one eight-year-old had difficulty in understanding a question in mathematics which referred to the volume of a 3D shape because his interpretation of volume was 'loudness'.

What does the teacher hope to achieve?

Words should be part of an oral vocabulary before they are part of a reading vocabulary. In mathematics, many words are first met in the textbook, without them being part of the child's oral vocabulary. It is hoped that children will gain the ability to think and discuss mathematics confidently using words appropriately and precisely.

What does the teacher do?

Initial teaching
For each unit in the mathematics programme of work, mathematical vocabulary could be listed under headings of 'known words', 'words to be consolidated' and 'probably new words'. The 'known words' could be quickly revised, possibly through the use of flash cards or directly from the mathematics textbook (workbook or workcard) both for the ability to read them fluently and to explain what each means. 'Words to be consolidated' could involve a child in providing an explanation for which others in the group have to identify the appropriate

word. These words could also be the content of a guessing game where you provide clues and the children have to identify appropriate words from a list or extract until all are eliminated except one. For example:

> 'Look at these words: *triangle square seven height six breadth.* I am thinking about one of these words. My word tells us how long something is. Which words do this?
> Here is a second clue. My word tells us how long something is from the ground upwards. Which word am I thinking about?'

'New words' could be researched by individual children themselves. Then the children could attempt a pronunciation of each word as individual lists are amalgamated by the group. The list should be discussed by the group so that each word is understood and able to be read. Mathematical words could also be the content of writing lessons.

Encouraging 'reading' as part of the teaching/learning procedure
A repetitious routine where only a textbook is used for written mathematics and where new vocabulary is always identified for the children should be avoided. If questions are read to the children accompanied by an explanation, there is little challenge to the children to read and interpret for themselves. The approach to lessons could be varied. Sometimes 'reading' mathematics should be the focus of the task and the resource might be a child's book where shapes, numbers and/or measures feature prominently. Descriptions, explanations or instructions written by classmates could be a resource for reading. Illustrations, diagrams, tables, graphs and computer graphics could be resources to be interpreted and explained. Data in newspapers, magazines and documents, especially those relevant to a project, could be read aloud by one child to the others in the group.

Setting tasks for the children
Example tasks are:
- *Use of a mathematics dictionary*
  Work with a partner. Read what the mathematics dictionary states for 'rectangle' (or any concept). Talk about what a rectangle is. Agree about how to explain a rectangle to the others in the group.
- *Explaining what you read*
  Read this newspaper report (textbook or workbook page). Note any words you are not sure how to say aloud and any you don't know the meaning of. Discuss these words with me. Later you will read the report to the group and lead a discussion about it.
- *Interpreting a graph*
  Write about what this graph shows.
- *Working with a partner*
  Work with a partner to do the questions on the workcard. Read each question in turn. Agree what each question tells you to do. Work out the answer by yourself. Compare answers and discuss any which are not the same.
- *Explaining instructions for a game*
  Read the instructions for this calculator game and then explain how it is played to the others in the group.
- *New words*
  Find words on page 'x' of your mathematics textbook which you are not sure how to say or what they mean. Write each word on a blank flashcard for me to use with your group.

Coping with difficulties which might arise

Children who have difficulties with reading skills would require more practice than others. They could have the opportunity to work sometimes with a more able and patient classmate and sometimes with you as well as working on their own. Reading to carry out their mathematics gives a purpose to their reading and may increase motivation.

The children with 'mechanical' reading difficulties might be helped by:
- identifying the initial letter of the word
- realising the shape of a word – the flashcard can be cut around the upright strokes and the tails of the letters to emphasise this
- relating the words to pictures on a poster
- finding a specific word in an extract and marking it with a highlighter.

The children with difficulties of interpretation can sometimes be helped by:
- reading the extract aloud
- telling a classmate what a sentence, then paragraph, is about.

The children with difficulties in summarising might try:
- finding the most important word in a sentence, then in each sentence in a paragraph
- finding their own words to tell the main point of the sentence, then paragraph, and gradually reducing these to a minimum.

Record keeping
The children experiencing difficulties with mathematics vocabulary could have their own set of flashcards, or list of words, for daily practice. Words are excluded as a result of repeated success and new words substituted. A record of this can be kept by the child so that you are aware of the words covered and those being dealt with currently. For the other children, a comment on progress through the vocabulary noted in the programme of work may be all that is necessary.

What do the children do?

Just as when learning a foreign language, children develop a vocabulary for mathematics which:
- is understood when others use it
- is used orally
- can be read aloud
- can be read with understanding
- can be written.

NINE

Focus on children doing and recording

USING RESOURCES

What do we mean by 'using resources'?

Part of children's learning in mathematics is to be able to use equipment such as a ruler, a metre stick, a metric tape, and a trundle wheel for measuring lengths; a set square and a pair of compasses to draw 2D shapes; plastic shapes or straws to construct 3D shapes; scales of different types for weighing; measuring jugs and a displacement can to find capacities of containers. The children also use unstructured and structured materials to represent number and to gain an understanding of relationships such as 'is more than' and operations like subtraction.

unstructured like buttons

structured like base ten pieces

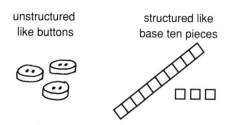

Visual aids, models, games and real life items such as groceries, maps, toys and clothing are among the many other resources used to help children learn mathematics with understanding. The calculator and the microcomputer are both important resources for mathematics learning in primary schools. They are felt to be so

important that each is dealt with under a separate heading in this section.

The teacher, of course, is the most important resource in the classroom.

What organisation is involved?

Organisation

Resources tend to be introduced by the teacher, or explored by the children themselves, often while working as a group. It is important that each individual has an opportunity to use or handle the resource as the group carry out different tasks. This may involve you in designating roles to the children, for example, 'Michael will read the instructions. Joy will use the balance. Mark will check what she does. Fiona will record the results.' The children can change roles until they have done each of them.

Location

Some resources require a specific location because of where they are kept, their size, or the need for electricity or water. For example, measuring jugs are left beside the sink, the toy cars on the floor, the video tape in the audio-visual room, and the trundle wheel in the playground. Others, like a map or game, may require a specific amount of table surface. Some such as a tape-recorder may be best used where the children can have silence and are not likely to disturb others.

Communication

Children should not only be able to use resources but also be able to describe them and/or explain their use to others.

Feedback

The children require to be observed occasionally, especially when they first begin to use a new piece of equipment. This will help to ensure that incorrect or inappropriate methods are not adopted which are difficult to unlearn.

What does the teacher hope to achieve?

Concepts can become meaningful and have relevance through using resources, for example, a child who has measured with a metre stick should gain a mental picture of the metre as a unit for measuring lengths of rooms, corridors, buildings and boundaries. Children who read a weight scale and worry about which mark the pointer is nearer, should gain some understanding about the approximate nature of measurements. Children often require experience to make many physical quantities meaningful, for example, objects must be handled before weight can be felt, and the force with which water can flow from an up-ended container often surprises children as does the area spilt water can cover. Children who have had experience with a range of equipment should develop strategies about how to use and care for it. This could give them confidence to handle resources at work and at home throughout their lives.

What does the teacher do?

Initial teaching

The procedure for using some pieces of equipment effectively needs a step-by-step explanation. For example, when children use a balance or set of scales they should be taught to:

- check if the surface is level

- consider if the equipment is appropriate for the task (for example, pans of the balance large enough, scales with an acceptable maximum weight)
- balance the empty pans or make sure the pointer is at zero
- make a decision about the balance or the pointer position by looking at it at eye level (usually horizontally, but for bathroom scales, vertically).

Encouraging 'using resources' as part of the teaching/learning procedure

If children are to be encouraged to use resources, they must be readily available and in working order. Part of the children's training in using resources could include:

- knowing where equipment and other items are kept in the classroom, in the open area or in a special resource room
- instructions about how to collect an item, for example place a flashcard with the name of the item on a special board indicating what has been collected and is in use
- telling the teacher if an item appears damaged or broken
- returning the item to its appropriate place and removing the flashcard.

More ordinary items like scissors, felt pens, cardboard can be available for class use, although many teachers find supplying such items to a group, where supply and care can be easily monitored, is more successful. Individual children could have the responsibility to use and replace some items.

Setting tasks for the children

Here are example tasks where the objective is to enable the children to use resources appropriately and effectively.

- Work as a group. Make a list of things you might use often in mathematics lessons such as scissors, card and rulers.
- Work as a group. Plan storage for the list of items for your group. Remember items should be easily found and kept tidy.

- Work as a group to construct the storage you planned.
- Explain to the others in the group how to use a trundle wheel (two metre sticks, the surveyor's tape etc.) to measure a length such as the corridor.
- Prepare, then say on the tape-recorder, instructions to use the bathroom scales (domestic scales, a spring balance, the displacement can).
- Work with a group. Plan how classmates should use textbook page x to help them learn more about . . . Try your plan out with a group.
- Work as a group. You are to make a plan/model of the classroom/games hall/school. List the materials you would like to use.
- Set up this video tape ready for use.
- Set the video recorder to record the programme.

Coping with difficulties which might arise
Some children may lack confidence in using equipment by themselves and may require more tasks where they work with someone else before they can take sole responsibility. Others may have difficulty in expressing instructions to use equipment and require a structure where they have guidance through numbered steps. Some may devise instructions as numbered steps for others. If giving the instructions orally instead of in written form is easier for some children, either let them work with another child who will write what is said or encourage them only to write one or two words rather than full sentences to remind them of each step.

Record keeping
The children should be aware of the skills they are being asked to achieve. Often they can keep a checklist of the equipment that they can use with confidence themselves. You could keep a record of what equipment and what routines for resource use have been used in the classroom with the different groups.

What do the children do?

Each child could be set the goal of learning to use equipment effectively. The child should also accept the responsibility for using resources with care and thought for other users.

USING A CALCULATOR

What do we mean by 'using a calculator'?

Using a calculator can motivate children to learn. They quickly find out that their initial thought 'that the calculator gives you the right answer' is only true when they press the correct calculator keys. Interpreting the question/problem, thinking what calculations have to be carried out, and determining the answer is a rewarding learning experience. Using the calculator to check mental or written calculations can be useful but is limited from both the interest and the learning involved.

What organisation is involved?

Organisation
This may well be determined by the number of calculators available for use. Ideally calculators should be available for individuals or pairs to use whenever they wish to do so.

Location
No special location is required for children using a calculator.

Communication
The children should be able to interpret what the display on the calculator shows and explain

this to others. They should also be able to describe the keys they use and their order for a calculation.

Feedback
The calculator provides some feedback to the child if certain keys are used which result in E appearing in the display. The child could be taught to check calculations either by repeating them or by finding the answer in another way, for example, by using the inverse operation. The child should also learn to evaluate the magnitude and the appropriateness of an answer in relation to the original question.

What does the teacher hope to achieve?

Children can:
- gain a wider experience of number
- focus on thinking through a procedure to find a solution rather than on the calculations
- try several procedures towards a solution without too much laborious calculating
- realise the benefit of using a calculator for work, home or leisure.

What does the teacher do?

Initial teaching
Discuss the purpose of using a calculator with the children but then leave them to find out for themselves how to carry out a specific procedure or operation. This task may, of course, be structured, for example by asking questions like:
- Is the calculator on? If not, find out what you do to put it on.
- When you press a number key, what happens?
- What do you do to add two numbers?
- What do you do to add four numbers?
- What do you do to subtract one number from the other?

You could provide a rule about when a calculator should be used. This could be 'at any time unless you have specifically been asked not to.'

Encouraging using a calculator as part of the teaching/learning procedure
Children can be encouraged to find answers in more than one way, with the calculator being regarded as one of the ways. A calculator should be considered by the children as a mathematical tool similar to a ruler – available and easy to use.

Some parents may be suspicious about ready access to a calculator, believing that this prevents the children from gaining 'paper and pencil' experience and expertise. This could be overcome by having a meeting with the parents where a workshop is run to show them that:
- the calculator has a value in carrying out different calculations quickly, but
- the person operating the calculator, not the calculator, is the decision maker.

Setting tasks for the children
Here are some sample tasks:
- What is the largest number you can show in the calculator?
- What is the smallest number you can show on the calculator?
- Add the prices marked on the items in this shopping bag. Check the answer by a different method.
- Multiply 37 by 46. Check your answer by doing a division.
- Divide 8449 by 17. Check your answer by doing a multiplication.
- Divide 68 by 7. What does the answer in the display mean?
- If there are 68 people to be taken to the theatre and a minibus can take 7 people, how many minibuses are required?
- If a container holds 7 litres of oil, how many can be filled from a drum of 68 litres and how much is left?
- What is the first number after 100 which divides exactly by 13?

- Continue this number sequence with appropriate numbers:

    2, 3, 5, 8, __ , __

- How can you enter $\frac{1}{4}$ into the calculator?
- What is one third as a decimal?
- Find 15% of £60 using the % sign on the calculator.
- Find $62\frac{3}{4}$% of £50. Explain how the calculator is used to find the answer.

Coping with difficulties which might arise

Many children are helped to achieve greater expertise and confidence in number by using a calculator, especially if the recall of facts is difficult for them. Some children could use the calculator for simple calculations so that they gain confidence in how to use it and to believe that it can produce the correct answer.

All children should be taught to consider the answer with relevance to what is asked in the question. Some children will benefit from estimating the answer by a mental approximation and then comparing the calculator answer with this.

Record keeping

You or the child could keep a checklist of success in working each of the basic operations with whole numbers and then with decimals. A checklist might also be made for specific techniques such as:

- rounding an answer to the nearest whole number
- finding further terms in a sequence
- checking by another operation
- making 'trials' to find a problem/puzzle solution.

What do the children do?

The children should learn to use the calculator effectively when it is useful to do so.

USING A MICROCOMPUTER

What do we mean by 'using a microcomputer'?

The microcomputer is used frequently in business whether it be the travel agent, the bank teller or the author, to store and retrieve information. The children should learn to accept the computer as a useful classroom resource which they use appropriately, effectively and with confidence over the full range of subjects in the primary curriculum. In this section we are particularly concerned with the use of the computer as a mathematics resource.

What organisation is involved?

Organisation

The computer can be used for a demonstration with a group. On most occasions, however, it will be used by an individual, a pair or a group, with a maximum of four children. Many teachers allocate different roles to the children working together at the computer and encourage them when working on their own to adopt this procedure. For example, there can be the typist or keyboard operator, the scribe or recorder, and the leader who allocates roles, leads discussions and manages the team effort.

Ideally every class should have at least one computer but where this is impossible, the school machines need to be timetabled. The children should know the arrangements so that they can collect and set up the computer for themselves and, when necessary, can plan their use of the computer.

Location

Where the computer is placed depends on the number of machines in the school and the layout of the building. It is often ideal to have several machines in the open areas rather than one in each classroom. Machines are best kept on a trolley so that they can be pushed along to different destinations. Simple rules are usually made up for permitted locations and means of conveyance.

Communication

The children could discuss what they are doing, or have done, on the computer as this encourages the use of mathematical vocabulary. Pupils can consider their input as the means they have of communicating with the computer and that it can communicate to them with a variety of prepared responses. However, they should also realise that this interaction has been devised by a human programmer and computers thinking for themselves are still an imaginary creation. They can be shown how to enter questions into a program to illustrate how a program's interaction is developed.

Feedback

The computer program can be prepared to give the child feedback on wrong answers, give a score based on difficulty and correctness, and sometimes question the child's input for comprehension and spelling. Children benefit from this instantaneous response. The teacher is often aware of computer feedback because of the noises made by the machine as it makes responses. Logged scores should be looked at and discussed with the children. It is also worthwhile to monitor a group at the computer to evaluate their use of it and their discussion.

What does the teacher hope to achieve?

The computer can be regarded as a 'helper' in the classroom. It interacts with the children gathered around it, allowing you the feedom to concentrate on the others. Good software can add further dimensions to an aspect of mathematics, for example:

- tutorial – adopts a teaching role
- drill and practice – provides examples, usually in small batches which are graded
- simulation – text and/or graphics describe a situation involving movement or a sequence of events, for example, the area of a river valley which is covered over time as the water floods
- demonstration – shows the child how to carry out a procedure such as counting, subtraction, drawing a shape, and finding the area of a composite shape
- problem solving – where the problem has several methods of solution and/or several solutions
- creativity – when the children use Logo or Dart commands to devise a program.
- ease of recording – where the computer draws, for example, different forms of graph, lines between named points to create shapes.

All of these possibilities allow you to use the computer for different purposes according to the mathematics being taught and the software available. Although the children can enjoy mathematical games on the computer for fun and general revision, much of the computer time will be integrated with other learning approaches to overcome specific objectives.

What does the teacher do?

Initial teaching

Your main task is to encourage children to be at ease with the computer. Some children may be tentative about using the machine while others will have the use of a machine at home and will be extremely confident.

As with other resources, the children could collect a specific disc, set this up in the machine and when they are finished turn off the computer and replace the disc in the storage system. A set procedure may be listed beside the computer and beside the printer to indicate the steps and their order for the initial switching on.

If the children are shown this once or twice, little more is usually required.

With most programs the children will simply follow the screen instructions. Sometimes, however, you may wish to use the computer screen as the 'blackboard' and gather the group around the screen to demonstrate, for example, drawing a path or shape using Logo commands.

If the children use a word processing or a Logo package, you may be needed to answer questions but if the basic commands are clearly set out on a card, the main task for the children will be to have enough hands-on experience.

Encouraging using the computer as part of the teaching/learning procedure
Availability of a machine, not only readily at hand but unmanned, is the main requirement if children are to be encouraged to select and use programs of their own choice. If children wish to write their own examples, text or program, this may require either a long session or a series of several short sessions to allow them to do so. You need to adopt a system where a child, or small group, can 'book' computer time. This is possible if you do not allocate the computer for a part of the day so that the children's own computer needs are met, and if you have time allocated on the work program for the children to have some choice in what they do. A rota will also be required where the children can organise their use of a printer.

Setting tasks for the children
Much of the work using the computer will be practice in using the keyboard and in following through programs. Here are a few examples of mathematical computer-based tasks:

- Write out steps telling classmates how to carry out the multiplication 36×47.
- Use the database program to store facts about the quadrilaterals (or numbers 1 to 20).
- Use Logo commands to copy this drawing:

- Use Logo commands to draw a picture.
- Use these two number games. Which do you think is the better game? Why?
- If you were going to write a computer program, what would you write it about? Why do you think that would make a good program?

Coping with difficulties which might arise
Some children might have poor manipulative skills and be frustrated in their slowness at using the keyboard. If the programs they use have limited typing and/or the typing is shared, their confidence should grow. However, the main solution will be to make sure they do not attempt too much and are therefore not rushed. Some children may have difficulties in using commands, for example, for Logo and for word-processing. You could structure the commands and the practice tasks so that the children learn and use only a few commands at a time.

Record keeping
If a timetable is kept for use of the computer, this can show:

- the names of the children who used it
- the period of time
- the program they used or the task attempted
- if the task has been completed
- if what was done was successful/correct.

The children could enter in these details and comments. This gives a useful record of each child's frequency at the computer and success in tasks. The teacher may like to write an informal weekly or monthly note about any pupils who are using the computer too little or too much, and plan time to monitor suspected weaknesses and praise noteable strengths.

What do the children do?

The children usually enjoy the variety provided by the computer for learning and recording.

PLAYING A GAME

What do we mean by 'playing a game'?

Games are ideal in mathematics for children to practise basic number facts, consolidate concepts and skills, and be involved in problem solving. The word 'game' seems to indicate competition and a winner, and this may be the criteria used to distinguish a game from an activity. This implies that a group of children work together and this is true in the majority of instances. However, sometimes an individual plays to improve his/her own score. The most popular games are usually those which have both an element of chance and of skill.

What organisation is involved?

Organisation
When children play in groups, you can form the group in different ways, for example by:
- asking one child to choose other players
- selecting a group of friends who respond well to each other
- selecting children from the same mathematics ability group
- selecting children by personality, for example, having dominant children play together.

Some teachers have games sessions where all the children are involved in playing games. Most teachers, however, build a game into either a group's programme of work or to a class session where each group is involved in a different activity. If a group has a free choice time, games can be used then too. In all these organisations usually only one, or at most two, groups are playing a game at any one time.

Storage of the games, with specific places for each of the game 'pieces' within the pack or box allows the children to be responsible for collecting and replacing them.

Location
The group, or the individual will require a suitable surface to play on. A game board should be made unmoveable, possibly by using blobs of Blu-tack on the underside corners. The players should be seated so that leaning across the board is minimised. The players need to talk to each other and so may be positioned in an open area, at the maths table, or in the games corner in the classroom so that they do not disturb others.

Communication
The game should be clearly explained to the players so that each is confident about the procedure and the rules, otherwise a player is likely to make a wrong move, perhaps be called stupid and then lash out at the others. The players should be expected to demonstrate good manners by:
- waiting for their turn patiently
- only moving their own counter or card
- replacing the dice in the container and handing it to the next player
- quietly offering help to another player where it is appropriate to do so.

Feedback
Satisfaction can come from the play. Children should realise that if they enjoyed playing, then it did not matter whether they won or not. Most children accept defeat gracefully when they realise that they were 'unlucky' today.

What does the teacher hope to achieve?

The children can be motivated to increase their memory and speed of recall of number facts. They can consolidate and extend their understanding of number, shape and measure. There are also social benefits as the children learn to play fairly and with thought for others.

What does the teacher do?

Initial teaching

Most games need a demonstration which may involve the teacher. When some children can play, they can show others what to do. It is a good idea to devise rules with all of the children for playing mathematics games and to display these as a reminder in the 'games corner'.

Encouraging 'playing a game' as part of the teaching/ learning procedure

Most children will not be aware that they are doing mathematics in some of their games so the children should be told what they might learn as they play the game.

Setting tasks for the children

Here are a few ideas for games:

- *Playday*

player board

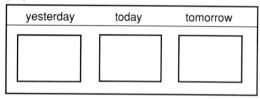

yesterday	today	tomorrow

cards

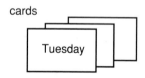

Tuesday

You need for two, three or four players:
- four sets of cards each with a day of the week written on it (28 cards altogether)
- a 'player board', as illustrated below, for each child.

Procedure:
- shuffle the cards and place face-down in a pile
- turn over the top card and place it face-up
- each player, in sequence, takes either the top face-down or the top face-up card
- the selected card is either placed on the player board or put on the face-up pile

- the aim is to complete the player board with an appropriate sequence of three days and the first to do so is the winner.

- *Playsum*

You need for two, three or four players:
- four sum cards for each player (they show an addition which has a total of 14, 15, 16, 17, or 18)
- total cards, that is cards which show a number in the range 12 to 20 (there should be five each of 14, 15, 16, 17 and 18, two each of 12, 13, 19 and 20, thirty-three cards altogether).

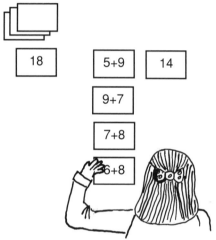

18		5+9	14

9+7

7+8

6+8

Procedure
- shuffle the sum cards and give four to each player
- shuffle the total cards and place these face-down in a pile
- turn the top card over and place it face-up
- each player, in sequence, takes the top face-down or the top face-up card
- the aim is to match all four sums with the correct total and the first to do so is the winner.

A more difficult addition, a subtraction and a multiplication version are all useful.

- *Playpay*

You need for two, three or four players:
- playing track as illustrated below
- plastic/card coins
- a dice in a shaker

– counters of different colours so that each player can have four of the same colour.

START	50p	25p	30p	free parking	10p
35p					30p
free parking					20p
30p					25p
25p					40p
50p	20p	10p	free parking	45p	5p

Procedure:
– each player is given five £1 coins
– each player, in sequence, rolls the dice and moves a counter around the board
– if a player lands in an unoccupied square, he may decide to park a counter there to collect money from other players who land there; the parking fee is £1
– each counter must begin at the start square and then move around the track in a clockwise direction covering as many circuits of the track as necessary
– a player must always move one of his/her counters when it is his/her turn
– if a parked counter moves no further, rent can be collected and another parking fee requires to be paid if the counter parks again in an unoccupied square
– if a player lands in an occupied square, he must pay the charge written on the square, unless the parked counter is his/her own
– the aim is to collect as much money as possible
– a player who cannot pay a charge has to retire from the game
– the winner is the richest player at a specified time.

● *So-low*
You need for the one player:
– a pack of playing cards with the picture cards removed (36 cards).
Procedure:
the player:
– shuffles the cards, then
– lays out the cards in rows of three
– counts the total of each row and removes the row of cards with the smallest total
– continues these steps until only three cards are left
– notes the final total
– plays the game (activity) another two times to find out if the total changes and by how much.

Coping with difficulties which might arise
Some children are uncertain about how to count along a track. They should be shown how to count the following square as 'one' and so on. Some children cannot hold a 'hand' of cards. They should be allowed to erect a screen (possibly a shoe box) and lay out their cards so that they can see them but others can't. Children could be allowed to play with the games cards by themselves so that they can read the words, make the calculations etc. by themselves beforehand to gain some confidence. You may wish to play a game with some children to monitor their difficulties and/or progress. Behaviour difficulties could be dealt with by giving a warning and then withdrawing the child from play.

Record keeping
No records need to be kept. A teacher may like to let a copy of a game be taken home to encourage further practice, or let the child show the family what he/she can do.

What do the children do?

The children should find games a fun approach to learning. Being a good loser and playing fairly could be socially beneficial for a child.

WRITING PRACTICE EXAMPLES

What do we mean by 'writing practice examples'?

Where children are shown a procedure or algorithm, they require practice to:

- become familiar with the steps in the procedure
- develop an understanding of what they are doing
- develop a language to explain what they are doing.

Most mathematics texts provide examples for children to write. These tend to be related to specific objectives, for example, adding tens and units without 'carrying', subtracting tens and units where there is an exchange of one ten for ten units, or multiplication of tens and units by a single digit. Sometimes there are examples drawing on a range of procedures previously taught. These are often referred to as 'check-ups' or 'tests'. Sometimes there are examples where the learner is required to state the operation to be carried out rather than perform the calculation. Many examples are presented as numerals and operation signs. Others are stated as word problems.

What organisation is involved?

Organisation

Written examples tend to be given as individual work. Occasionally children are allowed, or requested, to work in pairs. The same set of examples is usually given to all the children in a group. These may be specified as part of the child's programme of work where the individual finds the stated page, heads and dates a page in an exercise book, and proceeds to work through the examples at his or her own pace.

Location

The children tend to work at their own table/desk.

Communication

The children may be told which set of examples to do, but it is more usual for this to be written as part of group instructions.

Feedback

The completed examples may be kept by the child until the teacher comes along to ask about them, they may be placed on a correction pile, or may be taken by the child to the teacher's desk, standing in a queue to await attention and comment. Most teachers see this last method as unsatisfactory both for the children and themselves. They believe that work should be corrected when the child is present to discuss errors, so they plan correction time with the group. On some occasions the children will correct their own examples. If they do this, there should be a stated procedure for mistakes, for example, use a code to show if it was:

- a digit wrongly copied
- a wrong fact used
- a mistake in carrying
- the wrong operation used.

There could also be a rule about the number which should be correct to be regarded as a successful task. A child may set his or her own goal for the number which should be correct. The child should know what action to take, for example, rewriting the example, talking with a classmate, or talking with the teacher. You can monitor such correction by looking at a sample of exercise books.

What does the teacher hope to achieve?

The children should:

- have time to think through the procedure and develop their understanding
- gain confidence in carrying out a procedure

- produce their own record of growing expertise or specific difficulties
- gain mastery and speed in the use of basic facts and calculation skills
- develop and/or practise a strategy to interpret word problems.

What does the teacher do?

Initial teaching

It is worthwhile discussing the textbook with the children. They should be aware of the different sections of work, the position of the page title and any icons that are used to indicate features. The children should also be taught to scan the page to identify words which they do not know or which are less familiar. There should be a class/group procedure to find out more about such vocabulary, for example, ask a classmate, look up in a mathematical dictionary or ask the teacher. Flashcards and/or a class mathematics dictionary might be built up for such words by the children themselves following a procedure set by the teacher.

The children should have clear instructions about:

- where to write examples, paper or exercise book
- how to set out the date and a reference for the textbook page and the set of examples
- how to set out examples on the page
- how long they have to try to complete the examples
- what they are required to do if all the set work has not been completed
- where to place completed work
- when they will discuss this with the teacher as a group or as an individual.

Encouraging 'writing examples' as part of the teaching/learning procedure

Most teachers and children take it for granted that they will carry out written work as part of their mathematics. It is useful to encourage children when carrying out practical work or

problem solving to keep a record of calculations so that these can be checked. Children can be asked to write sets of examples for classmates to do. This should lead them to consider the individual steps in examples and how these can be made more complex to grade examples.

Setting tasks for the children

It is sensible not to set too many examples to be done at a time. Children benefit from tasks where they feel a sense of achievement. It is better to ask a child to carry out five calculations with the aim of setting these out with a good standard of presentation and getting them all correct than giving children twenty-five or more examples where they can be overwhelmed with the thought of the required effort even before they begin. It is important too that if a child does not understand or is carrying out the algorithm incorrectly, that this is confined to a few examples. It is also a waste of a child's time if the work is carried out successfully to do an enormous number of the same type of examples. You should avoid a reward for all correct examples of more to do. If further practice work is necessary it might be done using a software program on the computer or by playing a game.

Coping with difficulties which might arise

Some children find writing numerals laborious. It might be helpful to separate the examples into two types, some where they are required only to write the answers and others where they write the whole calculation.

Some children write crutch figures untidily and then become confused. Guidance on the size and position of crutch figures should help them. Children could discuss examples which are wrong so that they realise what was done incorrectly or was misunderstood. They might then do the calculations which were wrong again or you could ask them to take on the challenge of a few examples specially devised by you to focus on their particular difficulty. Children should be aware if their difficulties have been overcome or whether more practice will be required.

Record keeping

Children could use exercise books as a record of their work. Younger children can be asked to select an appropriate gummed saying or symbol for the standard of work they have achieved. Older children can look back over a week's work and write their own comment on what they have done. Some might keep a record of their best work in a folder, while others might like to graph a percentage score and then comment on this weekly or monthly.

You should note particular difficulties so that future work is planned for the children to overcome these.

What do the children do?

The children should try to achieve the objectives set for them regarding the standard of setting out work, the accuracy of the answers and the time taken for the examples to be completed. This is likely to mean the need for devising group or individual objectives for the children. The children could gain a sense of pride in their efforts as they are trying to improve on their own previous work.

USING A WORKCARD, WORKSHEET, WORK/TEXTBOOK

What do we mean by 'using a workcard, worksheet, workbook or textbook'?

Through the use of such resources, the children's learning is directed by the teacher without his or her being there.

A workcard is a non-expendable set of instructions and/or questions. The user is led through step by step or example by example. The materials which are required for a practical task are usually stated. The format for writing answers is usually indicated.

A worksheet is expendable and children are expected to write answers in the spaces or on the lines provided. Diagrams and tables are already drawn. Written work is minimised. A good model is provided for the children when they carry out recording for themselves in other tasks.

A workbook is simply a collection of collated and ordered worksheets. They are particularly useful for younger children and children with

learning difficulties. However, the recurring cost tends to make their use expensive whether they are produced by the teacher or commercially.

A textbook could be thought of as a collection of collated and ordered workcards but it is more than that as it tends to have notes, worked examples and illustrations to aid learning.

What organisation is involved?

Organisation

These resources are mostly used by the individual. However, the teacher can ask pairs of children or a small group to work together. This may be done to allow one child to help another with reading and/or recording. It may also be done to encourage children to talk about their mathematics work. A workcard can be written specifically for a group, particularly if practical work is involved.

To use these resources meaningfully as part of a teaching programme, you cannot usually allow children to work through cards/sheets/pages at their own pace. Each sequence of cards and page needs teacher input and teacher follow-up. Most teachers can only manage this if the class is working in two, three or four groups. There are likely to be a few working on individual programmes in a class but this is likely to be limited so that the teacher's time is used effectively.

Teachers usually find that it is best to have only workcards and worksheets available which the children are currently using or are about to use. This means the children can be motivated when the present supply is put away and a new batch substituted. The children should know when they are to use a card or sheet, where to find it, and where to place work for correction.

Location
No special location is likely to be required unless practical work is being carried out.

Communication
The children should have already met the key words so that they can attempt the interpretation of the instructions and questions for themselves.

Feedback
Ideally the children's work is corrected soon after they have finished the task. This may be carried out with a group and sometimes with individuals. The children can correct some of the work themselves using an answer book. If this does happen, you should select a random sample to see how well the task has been carried out. Children should be encouraged to produce their best work and recording. You may decide not to accept untidy work if this is likely to raise children's standard of presentation. However, you should also praise work where children have obviously made an effort to do their best.

It is sensible not to ask children to redo a task unless they have had further teaching to enable them to do it correctly.

What does the teacher hope to achieve?

The teacher hopes to:
- find out if the children can apply what they have been taught
- increase the children's mathematical reading vocabulary
- increase the children's ability to interpret written questions and instructions
- structure the children's time while teaching can take place with another group.

What does the teacher do?

Initial teaching
You need to teach the concept, the skill and/or the vocabulary before the children attempt work on the aspect of mathematics for themselves. Structured tasks should be regarded as follow-up to teaching, and not regarded as a substitute for teaching.

Encouraging using a workcard, worksheet, workbook or textbook as part of the teaching/learning procedure
Currently, commercial materials are usually well set out, attractive and colourful so children enjoy using them. Make sure your own products are attractive too. Use cut-outs from magazines or drawings and black felt pens, especially italic ones, to make cards and sheets look good.

Setting tasks for the children
Here is an example of a workcard produced by a first year student teacher. The task was to produce a workcard on length for children aged about eight or nine years old. The card includes too wide a range of tasks and uses rather too many pieces of equipment – a ruler, a metre stick and a tape measure. The use of a handspan seems inappropriate here.

Length Workcard

① Using a ruler, measure the length of
   a) your jotter   b) your foot   c) your pencil

② Using your handspan, measure the length of
   a) your desk   b) your ruler   c) this workcard.

③ Using a metre stick, choose six children in your
   class and measure their height.

④ Copy the table below and using a tape measure
   find the following

All About Me !

My height is ......
My leg length is ......
My arm length is ....
My waist is ....
My pace measure is .....
The length of my shoe is ....

Coping with difficulties which might arise

The children should know what procedure to adopt if they cannot read or cannot understand what they are required to do. Some teachers appoint a group leader to give help. Others have a series of steps for children to try, for example:

- pick out any word you don't know
- ask the group leader or your neighbour about the word
- read the question aloud to yourself
- put what you think you have to do into your own words
- check with a neighbour
- have a try
- if you are really stuck, go on to the next question and ask the teacher when he or she comes to your group.

Record keeping

Most children are encouraged to keep a record of work attempted on a grid or checklist. You may like to use a code to show that:

- the work has been looked at
- the work has been gone over with the child/children
- the child has grasped the main teaching points
- the child knows the key vocabulary
- the child has experienced difficulty
- the child needs extra practice
- the child has little or no understanding of the concepts and skills involved
- the child's recording is of a high standard
- the child should be able to produce a higher standard of recording
- this work has been checked at a later date and a good understanding maintained.

What do the children do?

The children have the opportunity to show:

- understanding of what has been taught
- neat and accurate recorded responses
- a checklist of work attempted.

RECORDING MATHEMATICS

What do we mean by 'recording mathematics'?

The phrase 'recording mathematics' is used to identify the learning approach which involves children in drawing diagrams, graphs and other forms of pictorial representation, devising tables, drawing shapes and patterns, plans and maps. Writing a story, a description, or an opinion about some aspect of mathematics is also included here.

What organisation is involved?

Organisation
The children may work as a group to record for a wall display, or work in pairs or individually on a smaller scale. Some of the tasks will result from solving a problem or working at an investigation. Others might result from talking with classmates. Some tasks could be teacher and/or workcard directed. In a few instances all the individuals in a group, or in the class, may attempt the same task. However, it seems more likely that the children will work on different tasks so that they believe they are producing something to show classmates.

Location
For group tasks an area of surface may be required, otherwise the children are likely to work at their desks.

Communication
Regardless of whether the task is carried out by the group or individuals, there should be the opportunity for the children to explain to others what they are doing or have done.

Feedback
You should try to find time to discuss the children's work with them, either as a group or as a class. The emphasis should be on the effectiveness of the record so classmates may want to comment too. Such comment should include praise as well as constructive criticism.

What does the teacher hope to achieve?

You might wish the children to achieve aims such as the ability to:
- express mathematical ideas in words and pictorial representation
- benefit from sketching diagrams or making notes while problem solving
- draw diagrams to illustrate steps in a task involving the use of materials.

What does the teacher do?

Initial teaching
You could demonstrate how a drawing, table or graph can help convey an idea, often more effectively than words. The blackboard, whiteboard or a large sheet of paper could be used as a background for bold diagrams. Some children try to work on too small a scale with a crowded layout and need advice to be bolder and clearer in their display. Children could be shown how to create and use conventional formats like the Tree, Carroll and Venn Diagrams but also to have scope for devising new forms. The children will gradually learn about different types of graphs and be able to choose an appropriate type for the data they wish to show. The children could discuss tables to identify key features and how these are recorded in rows and in columns.

Encouraging 'recording mathematics' as part of the teaching/learning procedure
You will need to use diagrams, tables and graphs to emphasise their usefulness. It seems likely if children are helped to understand something through a diagram that they will try using this means to aid their explanations to others. A supply of plain and squared paper should be available and a range of pencils and pens with different point thicknesses would be useful.

Setting tasks for the children
Here are a few examples of possible tasks:
For a group:
- Draw pictures to show what Charlie the cat might do in a day.
- Make up a table to show who in your group can play a musical instrument and which instrument.
- Design a wallpaper pattern using triangles.
- Make up a book which six-year-olds could use to learn words about length.

For an individual:
- Write what a diagonal is – use diagrams to help you explain.

- Sketch a graph to show how you have grown taller as you became older.
- Make up a table to show what items you have in your desk and in your schoolbag at the moment.
- Draw a picture of a tree which you can see from the window, then write a list of 'length' words which you could use to describe parts of the tree.
- Make up an entry for the class mathematics dictionary for the word 'pyramid'.

These tasks can be adapted to different situations and age groups.

Coping with difficulties which might arise
If children have difficulty presenting information in a specific form, they need advice about what is not clear/misleading/wrongly placed. They are likely to benefit from trying to interpret and analyse examples made up by the teacher and classmates.

Record keeping
Both individual and group work can be displayed for others to read and interpret. Questions may be posed by the creator or the teacher to encourage the children to focus on particular aspects of the presentation. Examples of individual work can be kept in a child's folder of work. Looking at several examples of their own work could help the children to have a sense of progress in their presentation skills.

What do the children do?

Mathematics can be dominated by children spending most of their time writing. They can be asked to record something when they don't see why this is necessary. Using this recording effectively for learning should mean that children realise the value of making diagrams, tables and notes to help carry out a task which focuses on explaining, describing and displaying.

PROBLEM SOLVING

What do we mean by 'problem solving'?

This approach involves children investigating and solving puzzles and problems. Children might be required to:
- interpret instructions
- agree with others about what the instructions mean
- select what knowledge they have which might be useful to find a solution
- apply knowledge to a new situation
- use new materials or familiar materials in a different way
- learn something new
- explain a possible solution to classmates
- listen to classmates and comment on what they have to say
- report on action taken and/or a solution.

Arithmetic has traditionally included word problems where children have to translate word questions into calculations. This is not the type of problem which is referred to here. A problem is an unfamiliar situation or one where the children have not previously found a solution.

What organisation is involved?

Organisation
The children may attempt problem solving on their own but teachers usually find that young children benefit from working in pairs and

older children from working in threes or fours. A successful organisation is to have four or five problems written out on cards, then ask a child to select a problem and choose one or two classmates to help him/her solve it. Young children, of course, have the problem read to them.

You can group children for a range of different reasons and it is important that the group learn to work well together if they do not already do so. You may need to teach social and working behaviours for this to happen.

Some teachers involve children in problem solving:

- as part of their programme of work
- as one approach to learning a subject like mathematics
- as part of integrated curriculum work
- as part of problem solving (giving it subject status with a timetabled slot).

Teachers differ as to how often they like children to carry out problem solving. It can be done every day, once a week, or once a fortnight. It often happens that initially problem solving is a separate subject attempted once or twice a week and then gradually, as the teacher and the children become used to this approach to teaching and learning, problem solving becomes an activity for one group during subject and/or integrated work.

Location

If the children are working as a group, they require a place where they have space to work and the ability to talk together without being disturbed or disturbing others. This may be possible in a corner of the classroom, in a walk-in cupboard, outside the classroom in the corridor, in the open area, or in a resources room.

Some problems can only be tried by other groups if they have not watched others find a solution. Others are open-ended enough to encourage each group to find a different solution.

Obviously these aspects affect the location. Ideally you should be able to monitor what is happening without participating.

Communication

Problem solving is an ideal focus for children to learn to share a task and to interact meaningfully with each other. Children could learn to identify and to use each other's expertise to advantage, to not just listen to each other but learn to modify the ideas of others, to realise when to push their viewpoint and when to give way to others.

Feedback

For many problems the solutions from several groups could be accepted and then discussed. Which solution is the most practical, the most attractive, the most interesting, and the most creative could be determined. In what ways would you change what you did now that the problem has been discussed with others. Some groups may want to try the same problem again and it would be up to you to decide if this was worthwhile.

What does the teacher hope to achieve?

You may wish to encourage children to:

- accept a challenge which at first might seem difficult to them
- have the confidence to guess and try to find a solution
- realise that you can fail to find a solution without being a failure
- interact effectively with classmates without teacher intervention.

What does the teacher do?

Initial teaching

Children need to learn how to attempt problems and how to work cooperatively with others. You might list behaviours which help the children to work together, for example:

- getting started by agreeing what the task is about
- working together by:
  - expressing opinions or suggestions
  - listening to what others have to say
  - encouraging others to express their opinion
  - sharing any practical or recorded work
- reporting to others what you did and your solution.

Each behaviour could be discussed with the children and a goal set for them to try to implement the behaviour when they are solving problems. When the children seem to be adopting the behaviour consistently, another is focused on.

You could concentrate on 'guess and check' as a strategy for younger children to use. The older children can extend their approaches to include strategies such as:

- draw a diagram/make a model
- make a list/make a table
- look for a pattern
- simplify
- eliminate.

Each strategy has to be discussed with the children and then suitable problems provided to try this way of working. The teaching of behaviours and strategies should be kept separate from the children being engaged in problem solving. When you are teaching or using a 'practice' problem, you will be interacting with the children. When the children are solving problems, you should not participate as the children must interpret the instructions and decide for themselves what to do. Your role is to monitor progress and discuss the children's solution with them.

Encouraging 'problem solving' as part of the teaching/ learning procedure
You could discuss problems which arise in the classroom with the children. This could demonstrate that life is full of problems and

people need to find a method of trying to find a solution. It could also let children see that sometimes a problem has to be left and returned to later or that sometimes a solution is not found. Class management problems can be used, for example, 'how to dissuade children from running in the corridor'. Choice decisions can be considered, for example, 'What novel will we read for the last five minutes each day?' A lack of information can be tackled together, for example, 'What is this pebble made of?'

You should find situations where the children can learn through problem solving rather than by being led through a task by using a workcard or worksheet. Ideally, problems are expressed in one or two sentences. They should interest the children so that they want to do them. Resources should be readily available for the children to use. A large box of junk materials is particularly useful. Younger children may be given some materials to help them get started. Older children could collect their own materials unless a specific item is mentioned in the instructions and would be supplied with the problem card.

You should show interest in the children's solutions and give praise for the good behaviours and strategies used by the group.

Setting tasks for the children
Here are just a few examples. Problems which are written for younger children can sometimes be used for older children, but problems for older children are most unlikely to be suitable for younger children.

For infant children (aged 5–7) – the instructions may need to be read to the pair or group of children and materials for the problem given to them.

- Make paper slippers which will fit everyone in your group.
- Make a parcel which will balance (weighs about the same as) this one.
- Write 2 so that it fills a square. Write the same kind of 2 to fill the other sizes of squares.

For juniors (aged 7–9) – they should be able to read the instructions and collect most of the materials for themselves.

- Decide what is the largest number you can write.
- Make four different nets of a cube.
- Make a Plasticine 3D shape which has three faces.
- Make a timetable for tomorrow for a new pupil coming to your group.
- Take the measurements of one of you for a jumper. List these to send to Gran.

For seniors (aged 9–12) – they should read the instructions and collect their own materials.

- Make up a game to give classmates practice with decimal fractions.
- Make up a calculator game that one person could play.
- Complete the number pattern 2, 4, —, —, —, — . . . in three different ways.
- Make a paper pattern for a suit of armour.

Coping with difficulties which might arise
Some children find it difficult to think of ideas. A brain-storming session might help, for example, produce an item like a sheet of paper and ask the children to think of as many things as possible that it could be used for. Other children find it hard to work cooperatively. It is useful to designate roles if the children cannot allocate these themselves, for example, leader, recorder, maker, reporter etc.

Record keeping
A note of children who have a difficulty and the behaviour which they are not achieving is useful. Every so often you may like to gather a group to discuss a specific behaviour with them. The children may be able to explain what they find difficult and you could suggest improved/alternative ways of working.

What do the children do?

Children could learn to:
- interpret instructions more effectively
- modify their behaviour to achieve the best group solution
- evaluate what their own group, and other groups, do constructively
- gain confidence in solving problems.

    Most children get tremendous satisfaction from solving a problem with a solution that they know is acceptable without having to ask.

SECTION THREE
SAMPLE LESSON NOTES

T E N

Introduction

Section One has been concerned with the content which is taught in primary schools and Section Two has suggested a range of approaches for teaching and learning this content. This section provides sample lesson notes to illustrate how specific content might be used for a particular teaching and learning approach.

The sample lesson notes are intended to give you a structure which can be used or modified when planning your own lessons.

The format for each lesson uses the following headings:
- Approximate age group
- Attainment targets for the National Curriculum in England and Wales, in Northern Ireland and for the 5–14 Guidelines in Scotland
- Context within a theme
- Organisation
- Resources
- Key language
- Aims
- Objectives
- Procedure.

The attainment outcomes for England and Wales are referenced, for example,

as: **Ma 5, 1a** where **Ma** indicates a mathematics target; **5** states the Attainment Target; **1** gives the Level; and **a** refers to the Outcome.

The attainment targets for Northern Ireland are also included.

The attainment statements for the 5–14 Guidelines in Scotland are not numbered, so a reference system has been devised, for example, as: **IH Organise B2b** where **IH** indicates the Attainment Outcome; (Information Handling); **Organise** is the name of the Strand; **B** gives the Level; **2** gives the Attainment Target; and **b** gives the subtarget.

Most of the lesson notes are for a single session, sometimes with a class but mainly with a group. One set of lesson notes suggest work for three groups of different ability. The work is on the same aspect of mathematics but differentiated to meet the needs of the different abilities. Examples have not been given where the groups are tackling different work, either in mathematics or in a range of subjects at the same time. Two sets of lessons are suggested to show development of some work with a group.

ELEVEN

Lessons for 5–7-year-olds

SORTING FRUIT

APPROXIMATE AGE GROUP
5-YEAR-OLDS

*National Curriculum
and 5–14 Targets*

England and Wales	Ma 5, 1a
Northern Ireland	D1a
Scotland	IH Organise A3
Context	Fruit
within themes such as	Shopping or food
Organisation	Group talking with the teacher

Resources

- A variety of fruit, for example, apples, pears, oranges, bananas, grapes, plums. One of each fruit forms a set. Several sets are required.
- A knife (round-ended for safety).
- Pairs of trays, plates or hoops.

Key language

Sort, fruit, skin, seeds, green.

Aims

To encourage children to identify attributes of objects for sorting and to discuss these.

Objectives

Children should:
- identify different attributes of fruit
- describe fruit
- sort fruit
- describe the sets in a sorting of fruit.

Procedure

1 Collect one piece of each fruit. Ask the children about each piece, encouraging them to describe the colour of the skin, the name and whether they like to eat it. Halve each so that the children can see if it has a stone, hard seeds or no seeds. The halved fruit should be set aside for the children to refer to when they are sorting.
2 Use one whole piece of fruit and two plates. Place all the fruit with a green skin on one plate and all the other pieces on the other. Explain to the children that you have sorted the fruit. Ask them to tell you what they can see about the fruit on this plate (the one with the green coloured fruit). Tell them the fruit on the other plate can be called 'not green' or

'does not have a green skin' (the negative of the description given to the first plate).

3 Whisper to a child to sort the fruit with those he/she likes on one plate and those not liked on the other. Ask the other children to try to say why they think the fruit has been sorted like this. Encourage answers by saying that the fruit does not have to be sorted because of the colour or anything that they can see. Let the child who did the sorting explain it. Emphasise that the second plate is described as 'fruit Michael does *not* like'.

4 Give each pair of children two plates and a selection of fruit. Tell them they are to sort the fruit *in a different way*, not by colour or because they like them. Tell them you wonder if they can find a way no one else will think about. Leave them to discuss and sort.

5 If a pair find an immediate sorting, provide a second set of fruit and plates with the challenge of finding yet another different way.

6 Gather the children around each sorting and encourage them to guess the reason for the sorting. The 'sorters' can comment on the responses. Discuss where the criteria is a matter of choice. Talk, where necessary, about how a sorting can be improved. The children may sort for criteria such as:

- has a skin which you peel before eating
- has one stone inside
- has small seeds inside
- is very juicy
- has a thick skin
- has a thin skin
- we have at home
- Gran likes
- feels soft
- feels hard.

HOW MANY?

APPROXIMATE AGE GROUP
5-YEAR-OLDS

National Curriculum and 5–14 Targets	
England and Wales	Ma 2, 1a Ma 1, 1b
Northern Ireland	N1a
Scotland	NMM Range A, 1a
Context within themes such as	Painting Myself or almost any theme
Organisation	Group using resources

Resources

- A set of flashcards for 'how many?', 'one', 'two', 'three', 'four', 'as many' – you need more than one of each of the number names.
- Painting equipment, for example, newspapers, brushes, sheets of paper, tubs for water, blocks or tubs of paint.

Key language

How many, one, two, three, four, as many.

Aims

To give children the opportunity to recognise and say some number names.

Objectives

Children should:

- be able to make sets of one, two, three and four
- recognise sets of one, two, three and four
- read the number names one to four and the phrase 'how many'.

Procedure

1 Tell the children they are going to set out the 'painting table'. Ask them to arrange four sheets of newspaper and leave them to organise this.
2 Ask the children how many easels they have put up, then to explain how they know there are four. Encourage answers which refer to counting and answers which indicate the children can recognise four.
3 Gather around a table. Put out three sheets of paper. Ask the children to say how many there are without counting. Ask one child to count as a check.
4 Repeat the questions about recognition and the check by counting other items, for example, four brushes, three water tubs, two blocks of paint, one box of paints.
5 Pretend you have forgotten how many there were of each object and suggest we put the number name beside each set. Help the children to read each flashcard. As you do this pile the objects together, or put them in a box, so that they cannot be recognised or counted.
6 Show the children the flashcard for 'how many?'. Discuss each word and the question mark.
7 Go over the flashcards on each pile asking one child to read the 'how many?' flashcard and another to read the number name.
8 Show the children the flashcard for 'as many'. Explain that there are as many brushes as sheets of newspaper, that is one brush for each sheet, four brushes and four sheets of newspaper. Ask the children to put out as many water tubs and paint blocks as there are brushes. Ask the children to put the flashcard for the correct number name beside each set of things. Ask them to read the cards.
9 Finish by collecting a pile of flashcards, shuffling them and then asking a child to read each one.

POSITION AND MOVEMENT

APPROXIMATE AGE GROUP 5- OR 6-YEAR-OLDS

National Curriculum and 5–14 Targets

England and Wales	Ma 4, 1b Ma 1, 1b
Northern Ireland	S1d, 2c
Scotland	SPM Position A, b
Context within themes such as	Going along the street The street
Organisation	Group using resources

Resources

- A toy car and a street card for each pair of children.

The street card shows a square pattern. Where lines meet there should be a black or coloured square.

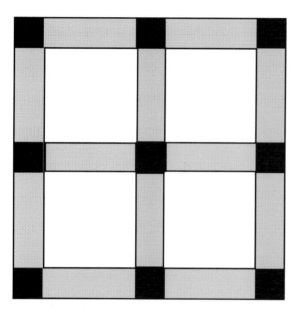

Key language

Right, left, turn, forward, back.

Aims

To give children the opportunity to follow or give instructions.

Objectives

Children should:

- know the directions right and left using themselves as reference
- move the toy car according to the instructions 'forward', 'back', 'turn to the right', and 'turn to the left'
- give instructions for a classmate to move the car.

Procedure

1 Arrange the group so that everyone is in one or two rows facing the same way. Ask the children to put their *right* hand out at the side. Now their *left* hand out at the side.

2 Tell them that people who travel on the road need to show others when they want to turn into another street. They can do this by putting out a hand. Tell the children to pretend they are on bikes and want to turn to the left. They should put out the correct hand. Repeat this with the children pretending to be bus drivers, lorry drivers, taxi drivers, and on a tractor.

3 Show the children the street card and explain that the grey lines are streets and the black boxes are stopping places.

4 Produce a toy car and place this on the middle coloured box. Ask one child to sit in a chair placed so that the child is facing the same way the driver of the car would be facing. The other children should watch from behind the seated child.

5 Give the child directions, for example: 'Go forward. Turn to the right.' or 'Turn to the left. Go forward.' Show the child how to put out the appropriate hand before turning, and after turning, *to rotate the street card* so that the child and the car are always facing the same way before making the next move. Forward and back moves are made from one black or coloured square to the next.

6 Give other children the opportunity to follow your directions. Then ask a child to give the directions to a classmate. Repeat this with different children. Those having difficulty could use 'bracelets' with right and left written on them.

7 Pair the children and provide a car and street card for each pair. They should work with the driver seated and the instructor standing alongside, with both facing in the same direction as the car. The children should change roles each time. You may want to pair

a child having difficulty with an able child or work for longer with such children yourself until they are confident about 'right' and 'left'.

8 The pairs may progress to choosing a different starting place, and/or giving a sequence of instructions.

ORDERING EVENTS

APPROXIMATE AGE GROUP
5- OR 6-YEAR-OLDS

National Curriculum and 5–14 Targets	
England and Wales	Ma 1, 1b Ma 4, 1c
Northern Ireland	M1
Scotland	NMM Time Aa
Context *within themes such as*	Spottie the dog Pets or animals
Organisation	Group talking with the teacher and classmates

Resources

- Two sets, each of five pictures showing Spottie the dog involved in a series of different actions, for example, eating from his dish, playing with a ball, sleeping. *There should be no obvious order.* The pictures on the next page could be used.
- Two sheets of paper, each with a row of 'windows' to accommodate the five pictures.

Key language

Before, after.

Aims

To give the children the opportunity to order a sequence of events and discuss these using the words 'before' and 'after'.

Objectives

Children should:
- be able to give an opinion about whether one event should occur before or after another
- agree with others about the order
- explain the chosen order
- answer questions involving the terms 'before' and 'after'.

Procedure

1 Lay out the pictures face up in a random arrangement.
2 Ask the children who is in all the pictures.
3 Ask a different child to talk about each of the things the dog, Spottie, is doing.
4 Show the children the sheet of paper with the row of 'windows'. Ask a child to place one of the pictures in the middle window/box. Ask another child to choose another picture and all of the children should agree whether Spottie did the activity it shows before or after the picture in the middle window. The children should be encouraged to say why they think Spottie did the activity before or after. The picture is then placed in the row according to the agreed decision.
5 Ask a child to select another picture and again, the group should discuss whether the activity should go before or after each of the pictures already positioned. This may involve moving the pictures already placed, so advice may be required about how to do this. Encourage all the children to say something.
6 Sub-divide the group into two smaller groups where individual participation may be increased. One group can work with the set of

pictures already being used. Place the two positioned pictures for the second group so that they carry out the same task. Leave the children to discuss and order their remaining pictures.

7 When they are ready, let the children tell you and members of the other group about their

chosen order. Question each group using the terms 'before' and 'after', for example, 'Did Spottie bury his bone before he went to sleep?', 'When did Spottie chase the cat?' and 'What did Spottie do after he ate his food?'

8 You may wish each group to tell their 'story' about Spottie to another group.

ONE QUARTER OF AN AMOUNT

APPROXIMATE AGE GROUP
6- OR 7-YEAR-OLDS

National Curriculum and 5–14 Targets

England and Wales	Ma 2, 2c, Ma 1, 2b
Northern Ireland	N2b, 3c
Scotland	NMM Fractions B
Context *within themes such as*	Quads The family, news
Organisation	Group talking with the teacher

Resources

- Four cut-out figures of children – these could be cut from a catalogue, be card silhouettes or figures made from pipe-cleaners.
- A cake or a circle representing a cake and a knife or scissors (round-ended for safety).
- Sets of interlocking cubes.
- Sets of items such as 8 apples, 12 pence.

Key language

Quarter, a quarter of.

Aims

To revise one quarter as one of four equal parts, present one quarter of an amount as one of four equal parts or shares and give children the opportunity to talk about the concept of a quarter in both situations.

Objectives

Children should:
- be able to describe one quarter of a whole

- realise that 'making quarters' means making four equal parts
- realise that one quarter is one of four equal shares
- be able to find one quarter of amounts less than 20
- describe the procedure used to find one quarter of an amount.

Procedure

1 Show the children the 'cake' and one of the cut-out figures. Tell them that the figure has to get one quarter of the cake. Ask how the cake can be cut into quarters.
2 Ask another child if the pieces are quarters and explain why or why not.
3 Establish with all the children that one quarter is one of four equal parts.
4 Introduce the four figures, explaining that they are quadruplets (or quads for short), that is children who are all born on the same day. Tell the children that the quads like to have equal parts of everything.
5 Ask a child to share the whole cake equally among the quads and to say what fraction each quad would get.
6 Show 4 cubes which are joined together to make a tower. Tell the children that the cubes have to be shared equally among the quads. Explain that each quad will get one quarter of the tower. Suggest that, like the cake, the tower is broken into four equal pieces. One piece, is given to each quad.
7 Ask what fraction of the tower and how many cubes a quad has? Tell the children they have found that: 'one quarter of four is one'.
8 Show 8 cubes joined together in a tower. Ask a child to say what fraction of the tower a quad is to get? Ask the child to describe how he/she finds one quarter of the tower.

Encourage a step where the four parts are checked to show that they are equal.

9 Ask the children to complete this sentence orally: 'One quarter of eight is . . .'.

10 Leave the children to work in pairs to find out practically one quarter of each of three towers (one of 16 cubes, one of 12 cubes and one of 20 cubes).

11 When you return ask different children to show you and their classmates what they did to find one quarter of each tower and what the numerical answer is.

12 Children who are finding this easy can be challenged to find one quarter of 8 apples and/or 12 pence.

ADDING THREE NUMBERS

APPROXIMATE AGE GROUP
6- OR 7-YEAR-OLDS

National Curriculum and 5–14 Targets

England and Wales	M2, 2a and 3a
Northern Ireland	N3d
Scotland	NMM Add and subt B, 1a
Context	An activity with number cards
within themes such as	Fun with numbers
Organisation	The class in pairs writing practice examples

Resources

• A sheet of paper and scissors for each pair of children.

Key language

Add, total.

Aims

To make practice examples of addition more fun for the children.

Objectives

Children should:
• be able to make a set of number cards
• write a horizontal addition for each row of cards
• find the total
• practise mental addition of three numbers.

Procedure

1 Give each pair of children a sheet of A4 paper. Demonstrate how to fold this into sixteen. Ask the children to cut along the fold lines to produce sixteen 'cards'. You may like to supply the cut pieces for some children.

2 Tell the children they are to write one number on each piece of paper. They are to use 0 to 9. How many of each number they use is their choice. (You may like to restrict the numbers to 0 to 5 for some children, remembering that the addition of three numbers is to be carried out mentally).

3 Tell the children they are going to make their own adding sums today. Take a set of sixteen number cards which you have prepared. If you wet the back of each it will stick to a vertical surface for a short time. Demonstrate how to lay out three cards in a row. Show the children how to copy the numbers horizontally with a plus sign between each

number and an equals sign at the end. Ask a child to give you the total of the three numbers. Repeat this demonstration for five rows, using all but one card which is laid aside. Ask the children which row has the smallest total? The cards in that row are laid aside, the others collected, shuffled and then laid out in four rows. Copy each row again and ask a child to find each total. When asked 'What happens now?', the children should realise that the cards in the row with the smallest total are set aside. This procedure continues until there is only one row to total.

4 Each pair of children should now try this setting out of cards, recording of numbers, finding totals and discarding a row. When the task is completed the children could compare their recording and answers.

5 The task may be repeated with a shuffled set of cards, and on this occasion the numbers do not require to be recorded. The children should check that they agree for each total.

LENGTH VOCABULARY (1)

<table>
<tr><td colspan="2">APPROXIMATE AGE GROUP
7-YEAR-OLDS</td></tr>
<tr><td colspan="2">National Curriculum and 5–14 Targets</td></tr>
<tr><td>England and Wales</td><td>Ma 2, 2d Ma 1, 2b</td></tr>
<tr><td>Northern Ireland</td><td>M1</td></tr>
<tr><td>Scotland</td><td>NMM Measure Ac, Ba</td></tr>
<tr><td>Context
within themes such as</td><td>Measuring
Our classroom</td></tr>
<tr><td>Organisation</td><td>Group devising and
answering questions</td></tr>
</table>

This is the first of two lessons on length.

Resources

- Strips of paper (some for words, some for sentences).
- A felt pen.

Key language

Long, short, tall, small, high, low, length, breadth, height, measurement, pace, metre.

Aims

To provide an opportunity for children to understand, say and read some length vocabulary.

Objectives

Children should:

- be able to devise some questions about length
- be able to say some length words
- be able to read some length words
- understand what each of these words means
- sort some length words.

Procedure

1 Tell the children that they are going to make up questions for classmates about measurements of the classroom. Write the word 'measurements' on a strip of paper. Ask a child to read the word and another to tell you what it means. Build up a meaning based on heights, lengths, how long, how high? As the children offer words and you supply some, write each on a separate piece of paper.

2 Go over the words asking children to read them. As each is read ask the children to make up a question to ask about the

classroom using the words long or high, for example, 'How long is the bookshelf?' 'How high is the light switch from the ground?' As the children offer questions and these are modified, where necessary, write each on a strip of paper.

3 Ask a child to read each question and explain what the 'length' word means. For example, 'How *wide* is the door?' means the measurement from here to here on the door'. Explanations like this from the children can be expanded by asking the children how they would answer the question, for example, 'about six footsteps' and/or 'about a metre'. These answers could help the children to reach a meaning for 'wide' as the number of units which make up *a straight line across the floor* (a horizontal length).

4 The children should find that several words can be explained in the same way, for example, wide, long, and length, can all be described as measurements of straight lines along the floor. This might encourage the children to introduce more words such as

breadth, and distance. Each word should be written and used by the children to devise a question about the classroom.

5 Words such as tall and height might be explained as 'a straight line from the floor up the wall'. High can be found to be different in meaning and be 'a place above the floor'.

6 Read through the words and encourage the children to think of some more, for example, by asking them to give you some opposites, for example, long—short, tall—small, high—low. They could also find 'comparing' words such as longer and smallest. These words could be used by the children to find more length questions about the classroom, for example, 'Is the desk shorter or taller than the chair-back?'

7 Some of the children could begin a display of a wordbank for length by arranging the words on a background sheet by using Blu-tack (these are likely to be rearranged so don't stick them permanently). The other children could display the questions on another sheet again using Blu-tack.

LENGTH VOCABULARY (2)

APPROXIMATE AGE GROUP
7-YEAR-OLDS

National Curriculum and 5–14 Targets

England and Wales	Ma 2, 2d Ma 1, 2b
Northern Ireland	M1, 2c, 3b
Scotland	NMM Measure Ac, Ba

Context **within themes such as**	Measuring Our classroom
Organisation	Group reading mathematics vocabulary

This is the second lesson on length.

Resources

- The words made up for the wordbank in the previous lessons.
- Strips of paper and felt pens.
- Textbook or workbook.

Key language

Long, short, tall, small, high, low, length, footstep, pace, metre, measure.

Aims

To provide an opportunity for children to understand, say and read some length vocabulary.

Objectives

Children should:
- be able to say some length words
- be able to read some length words
- understand what each of these words means
- sort some length words.

Procedure

1 Gather the children around the wordbank of length words. Remind the children that these are words about length. Ask children to read the words.

2 Tell the children that they are going to sort the words. Suggest that they begin with words which describe the length of things. Ask them to look at, for example, the door, and then to find words in the wordbank which could describe the door. Responses might be 'tall' and 'wide'. Ask about describing words for other objects until all the adjectives have been used. Include the comparative and superlative terms here too. As the children describe the objects they can be encouraged to find other length adjectives. All the adjectives should be rearranged into one part of the wordbank.

3 Tell the children that they could find words which are units of measure, for example, 'footsteps' and 'metres'. Ask them to find any

others in the wordbank. There are not likely to be many. The children might notice that some of the units are objects, parts of their bodies or some that they find on rulers. Ask the children to work with a partner to read one or two length pages of their mathematics books to find the names of more units. It is useful to have the textbook and workbooks which the children may have used in a previous year. Workcards too can be used so that each pair have the opportunity to find different words.

The children should keep a record of the unit names they find by writing each on a strip of paper.

4 The children should find units such as: straws, sticks, rods, handspans, palm widths, paces, and centimetres. The children's collection of unit names could be reduced so that there is a set in which all are different and these can be arranged together in another part of the wordbank.

5 Ask the children to read through some pages again to find length words which are not already in the wordbank. Again a record is made of each.

6 Work with the children to sort the remaining wordbank words and those they have just found. There could be 'name' words which are all like length – measurement, breadth, height, distance. There could be 'action' words which tell you what to do (verbs), for example, measure.

7 You may like one or two of the children to tell their classmates about the wordbank and the types of words they have found.

T W E L V E

Lessons for 7–8-year-olds

THE SAME TOTAL (GROUP 1)

APPROXIMATE AGE GROUP
7- AND 8-YEAR-OLDS

National Curriculum and 5–14 Targets

England and Wales	Ma 1, 3d
Northern Ireland	P2c
Scotland	PSE
Context	Puzzles
within themes such as	Problem solving
Organisation	Group solving a problem

You have decided to include differentiated problem solving as part of each group's programme of work. Each group will be sub-divided into twos or threes. These sub-groups will attempt the same problem which has been adapted for different abilities.

Each group will carry out the problem solving at a different part of their programme of work so that only one group is discussing in their sub-groups at any time.

Resources

- Squared paper.
- A copy of the instruction card for each sub-group.

Key language

Row, column, total.

Aims

To challenge the children to see relationships among numbers.

Objectives

Children should:
- be able to place numbers in boxes to obtain the same total in the row and in the column
- show that any five consecutive numbers can be used in the same way.

Procedure

1 Arrange which children will work together. This may be left to chance so that a group of three children come together as they complete all of the work programme, or you may allocate names to sub-groups.
2 Write 'problem solving' as the last item in the group's programme of work and allow about half an hour for the problem.
3 Have a copy of these problem instructions and squared paper available for each sub-group.

1 Place each of
the numbers 3,
4, 5, 6 and 7 in a
box so that the
row and the
column add to
give the same
total.

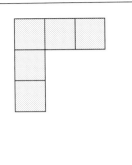

2 Find out if it seems to be true that any
five numbers which follow on after
each other, can be used in the same
way.

4 After each sub-group has been underway for
about five minutes, monitor the children's
understanding, if necessary, by asking them
what they are doing.

5 Plan a follow-up discussion with the class
where the groups can talk about what they
did and found.

THE SAME TOTAL (GROUP 2)

APPROXIMATE AGE GROUP
7- AND 8-YEAR-OLDS

National Curriculum and 5–14 Targets

England and Wales	Ma 1, 3d
Northern Ireland	P2c
Scotland	PSE

Context *within themes such as*	Puzzles Problem solving
Organisation	Group talking with the teacher, then solving a problem

Resources

- Paper and felt pen for a poster.
- Squared paper.
- A copy of the instruction card for each sub-group.

Key language

Row, column, total.

Aims

To challenge the children to see relationships
among numbers.

Objectives

Children should:
- adopt a 'starting' procedure for solving
problems
- be able to place numbers in boxes to obtain
the same total in the row and in the column
- find there are different solutions to the
problem.

Procedure

*This group should have problem solving as the second
task on their programme of work. It should have 'T'*

beside it to indicate the group will be working with the teacher.

1 Tell the children they are going to work in groups of three to solve a problem. Show them the instruction card and explain that what they have to do is written on a card like this. Ask them to tell you how they will begin.

2 The children should suggest reading the instructions. You could develop this to suggest that each child has a different role, for example, one is the reader, another is the thinker, while the third is the judge. The reader reads the instructions aloud. The children can discuss who will be chosen as the reader and what they should read, for example, all the instructions and then the first bit again. The thinker has the task of saying what he or she thinks they have to do. The judge comments on whether he or she agrees with the thinker and why or why not.

3 You might close the discussion by writing the starting procedure on a poster. For example:

To start a problem

- *read the instructions*
- *say what you think they mean*
- *agree about what they mean.*

4 Arrange which children will work together as sub-groups. Remind them about the reader, the thinker and the judge.

5 Have a copy of the following problem instructions and squared paper available for each sub-group.

1 Place each of the numbers 2, 3, 4, 5 and 6 in a box so that the row and the column add to give the same total.

2

You can do this more than one way.

Find out if Michael is correct.

6 Monitor if the sub-groups follow the suggested start procedure and their understanding about what they are to do.

7 Plan a follow-up discussion with the class where the groups can talk about what they did and found.

THE SAME TOTAL (GROUP 3)

APPROXIMATE AGE GROUP
7- AND 8-YEAR-OLDS

National Curriculum and 5–14 Targets

England and Wales	Ma 1, 3d
Northern Ireland	P2c
Scotland	PSE

Context	Puzzles
within themes such as	Problem solving
Organisation	Group solving a problem

Resources

- A copy of the instruction card.
- A pencil.
- Numbers on paper squares (to fit the boxes on the card) for each pair of children.

0	1	2	3	4

- Glue.

Key language

Row, column, total.

Aims

To challenge the children to see relationships among numbers.

Objectives

Children should:
- be able to add the numbers in a row and in a column
- be able to place numbers in boxes to obtain the same total in the row and in the column.

Procedure

1 Join Group 4 who have problem solving as the first task on their programme of work. Tell the children they are going to work in pairs to solve a problem. Arrange the children in twos and give each pair a copy of the instruction card. Explain that what they have to do is written on the card. Ask the children to follow the words as you read them.

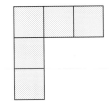

The numbers 1, 2, 3, 4 and 5 are in boxes so that the row and the column add to give the same total.

1 What is the row total?
2 What is the column total?

3 Are the totals the same?

4 Put the card numbers 0, 1, 2, 3, and 4 in these boxes to make the totals the same.

2 Ask the children to tell you what they are to do. Clarify any difficulties of understanding about the nature of the task.

3 Tell the children to write the answers to questions 1, 2 and 3 on the card and to use

the paper numbers for question 4. Do not supply the glue yet.

4 When the first pair are sure they have the numbers in the correct boxes, give them glue. This can be passed to the others when they are ready.

5 As each pair finish the problem they go on to the next item on their programme of work.

6 Plan a follow-up discussion with the class where the groups can talk about what they did and found.

TIME GAME

APPROXIMATE AGE GROUP
7- OR 8-YEAR-OLDS

National Curriculum and 5–14 Targets

England and Wales	Ma 2, 3e
Northern Ireland	M3b
Scotland	NMM Time Bb
Context	How I spend my time
within themes such as	My day or playing
Organisation	Group using a gameboard

Resources

- One or two copies of the illustrated gameboard.
- A time card for each player – this should be a piece of card about 3 cm by 2 cm showing a digital time such as 3:36, 3:48, 3:55, 4:02, 4:09, 4:12, etc. for example:

It is ideal if each time card is a different colour.
- A box of counters.
- A dice.

Key language

Later, earlier, minutes past, quarter past, half past, (quarter to may be used or forty-five minutes past).

Aims

To provide practice for children to compare times written in words, shown by digital display and on an analogue clock face.

Objectives

Children should:
- be able to read times given in words
- be able to interpret digital times
- be able to interpret analogue clock times
- understand the words 'earlier' and 'later'.

Procedure

1 Ask a child to tell you the time, for example, 'twelve minutes past ten'. Discuss with the children another way of saying this time for example, 'twelve ten'. Write different ways of showing the time, for example, in the words we have just said, using figures 10:12, using a combination of words and figures such as '12 minutes past 10', using a drawing of an analogue clock, and using the drawing of a digital clock.

2 Ask the children to each give you a time which is earlier than twelve minutes past ten.

Now ask them to each give you a time which is later than twelve minutes past ten. Discuss any difficulties which arise in their replies.

3 Take two of the game 'time cards' and ask which time is earlier. Take different pairs and repeat this question. Now take pairs of cards and ask the children to identify which time is later.

4 Give a child one time card and ask him or her to say if the time shown is later or earlier than the time on each of the other cards. Give another time card to another child and repeat the comparisons.

5 Place a blob of Blu-tack at each corner on the underside of the gameboard to keep it in place during play. Give time cards to two of the children. Discuss the rules which are written on the gameboard, for example:
Play for 10 minutes. We'll set the timer. When it rings we stop playing.
What is next? *Each choose one time card and take five counters out of the box.*

You need to remember which is your card.
Put it on any box on the track you want to. Continue so that the players learn how to move around the track according to the throw of the dice, how to compare times and to tell the other player the decision, give or take a counter. Play until the timer rings. Involve the other children by asking them if the player has said the time correctly, has made the correct decision about earlier or later, and had made the correct transaction with the counters.

6 Leave the children to play either as one or two groups (a maximum of four players is probably best).

7 These children should have the opportunity to play on another day to remind themselves of the rules. Then each could have the opportunity of showing another child or a small group how to play.

8 This game can be played by an individual too.

Early or late

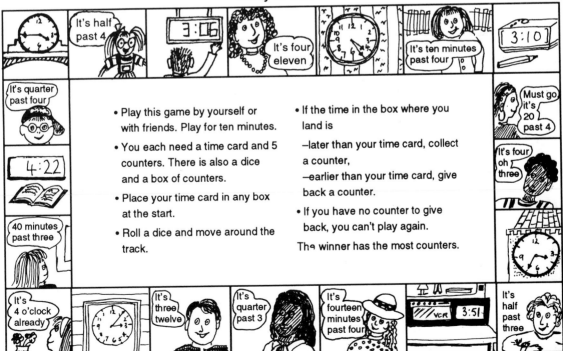

- Play this game by yourself or with friends. Play for ten minutes.
- You each need a time card and 5 counters. There is also a dice and a box of counters.
- Place your time card in any box at the start.
- Roll a dice and move around the track.

- If the time in the box where you land is
  –later than your time card, collect a counter,
  –earlier than your time card, give back a counter.
- If you have no counter to give back, you can't play again.
The winner has the most counters.

THIRTEEN

Lessons for 8–11-year-olds

INVESTIGATING THE SIX TIMES TABLE

APPROXIMATE AGE GROUP
8- TO 10-YEAR-OLDS

National Curriculum and 5–14 Targets

England and Wales	Ma 2, 4a Ma 1, 4d
Northern Ireland	A4b, c
Scotland	NMM Multiply C1a, D1a Patterns C
Context within themes such as	Multiplication Number patterns
Organisation	Class or group talking with the teacher

Resources

None required.

Key language

Multiple, digit, even, odd.

Aims

To consolidate the facts of the six times table and establish facts about the multiples of six.

Objectives

Children should:
- know how to build a multiplication table
- be able to explain the term multiple
- find a pattern in the multiples of six
- generalise about numbers divisible by three and by six.

Procedure

1 Discuss multiplication with the children and establish that it is a short way of calculating additions of the same number. Use examples such as:

   $1 + 1 + 1$ is the same as 3×1
   $2 + 2 + 2$ is the same as 3×2 and so on.

   Ask the pupils to write a few facts of the six times table like this. Conclude that we take a short-cut to adding by learning multiplication facts.

2 Ask a child to say the 'stations' of the three times table – 3, 6, 9, 12, 15, 18, 21 . . . Stop him or her and ask how you can find the next station if you have forgotten it. Discuss the pattern of adding three. Ask the children to find the pattern of the stations of the six times table.

3 Discuss how the stations are called multiples, that is they are made up of the same number

added to itself or zero. Multiples divide exactly by the number, for example, multiples of six divide exactly by six.

4 Ask the children to look carefully at the multiples of three: 3, 6, 9, 12, 15, 18, 21, 24, 27, 30. Point out that all the digits 0 to 9 appear in the units column although they are not in order. Ask them if they think there are more multiples of three and establish that there are: 33, 36, 39, 42, 45, 48, 51, 54, 57, 60 . . . and so on. Ask the children if they notice anything and establish that the unit digits are repeated again. Ask a child to predict other multiples and establish that the units digits will be repeated again and again.

5 Ask the children to investigate the pattern for the multiples of six:

6, 12, 18, 24, 30
36, 42, 48, 54, 60
66, 72, 78, 84, 90 . . .

They should find that the unit digits are repeated over and over again, but this time all the numbers are even. They may want to look again at the multiples of three and realise that they were both odd and even.

6 Leave the children to work in pairs to find a rule for recognising a multiple of three.

7 Someone should suggest that the digits add to make three, six or nine. The 'rule' should be tried out.

8 Leave the children to find a rule for recognising a multiple of six.

9 Someone should suggest it is the same as three. If this happens, ask them if fifteen is a multiple of six. They should realise that a multiple of six must be an even number in which the digits add to make three, six or nine.

10 You may want to challenge some children to investigate the multiples of nine to make a statement about:
   • the patterns of digits
   • the rule for recognising the multiples of nine.

NOTATION FOR DATES

APPROXIMATE AGE GROUP
9-YEAR-OLDS

National Curriculum and 5–14 Targets

England and Wales	Ma 3, 2e
Northern Ireland	P 3c, 4d
Scotland	NMM Time C 1d
Context	Letters
within themes such as	The Post Office, communications
Organisation	Group using resources

Resources

• A set of about six envelopes each with a different format of postmark, for example, the month written in words, Roman numerals, month before day as in the USA.
• Current calendars.

Key language

Most recent, before, after, order.

Aims

To give children the opportunity to order a sequence of postmarks.

Objectives

Children should:

- be able to interpret each postmark
- be able to write dates in different formats
- order the dates from earliest to most recent
- calculate each date of postage as a number of months and days before today.

Procedure

1 Have the envelopes ready in a random order.
2 Ask one child to write today's date large enough for all to see. Discuss what each part means.
3 Show the children one envelope and discuss the postmark – why it is used and what information it conveys.

This postmark tells the city and the district as well as the day, the month and the year of posting.

4 Leave the children to discuss the different formats of the other postmarks.

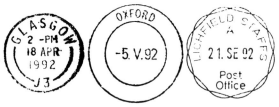

5 Ask the children to comment on their findings, for example, the different ways of recording the month in words, Roman numerals and Hindu-Arabic numerals.

6 Leave the children to draw postmarks showing today's date in these different ways.
7 Ask the children to write tomorrow's date in three different ways.
8 Ask the children what is meant by 'the most recent letter' and let them identify which of the given envelopes could have contained it.
9 Leave the children to order all the dates.
10 Using a calendar, discuss with the children a procedure to calculate how long it is since one of the envelopes was posted. For example:

- find today's date
- count back weeks by moving vertically or horizontally (depending on the calender layout)
- count back days to the date on the postmark.

This is a good time to remind the children that any seven days are one week, so a move from Wednesday back to the previous Wednesday is one week. When counting days, like moving along a track, count from the one next to the one you are in.
Some children will find it easier to cope with letters that are only a few days or at most three weeks old, while others will enjoy the challenge of moving back through several months.

11 Leave the children to calculate 'How long ago was each of the other letters posted?' The envelopes could be set out on a table and pairs of children, armed with a calendar, could move around these. Ideally have more envelopes than pairs of children so that there are always free envelopes for a pair to move to.
12 Ask the children to collect envelopes with different postmarks and bring them to school to make a display.

MAKING UP A WORD FORMULA

APPROXIMATE AGE GROUP
9- AND 10-YEAR-OLDS

National Curriculum and 5–14 Targets

England and Wales	Ma 3, 4a Ma 1, 4d
Northern Ireland	A 4d
Scotland	NMM Functions
	Level D
Context	Perimeters
within themes such as	Shapes
Organisation	Group talking with
	classmates

Resources

- Centimetre isometric paper.
- Centimetre squared paper.

Key language

Perimeter, regular shape, add, multiply.

Aims

To give children the opportunity to express how to find the perimeters of regular 2D shapes as a 'word formula'.

Objectives

Children should:
- know the meaning of the word perimeter
- calculate the perimeter of some regular shapes
- find words to state how to calculate the perimeter of one type of regular shape
- find words to state how to calculate the perimeter of any regular shape.

Procedure

1 Ask the children what the word perimeter

means. Arrive at a definition such as 'the boundary of a shape'.

2 Separate the group into two sub-groups.

3 Ask one sub-group to draw on isometric paper a pattern of equilateral triangles, the first with edges of 1 cm, the second with edges of 2 cm. They are to find the perimeter of each triangle.

4 Ask the other group to draw on square grid paper a pattern of squares, the first with edges of 1 cm, the second with edges of 2 cm. They are to find the perimeter of each square.

5 Return to the first group when they have calculated the perimeters of about five triangles and ask them to write in words how to find the perimeter of *any* equilateral triangle. They might write 'Add the three edges'.

6 Return to the second group and ask them to write in words how to find the perimeter of *any* square. They might write 'Multiply one edge by four'.

7 Ask both groups to discuss their statements with each other and then to agree about words which tell classmates how to find the perimeter of any regular 2D shape. They might decide on 'Multiply one edge by the number of edges the shape has'.

8 Give the word formula to another group to find the perimeter of a given regular pentagon, for example with an edge length of 4 cm.

9 Ask the devisers of the word formula and the users to talk together and to agree a final wording, for example, 'Find how many edges the regular shape has. Find the length of one edge. Multiply the length by the number of edges.'

10 The devisers could be asked to make a poster giving the word formula to show it applies to a range of regular shapes.

ESTIMATIONS OF WEIGHT

APPROXIMATE AGE GROUP
9- OR 10-YEAR-OLDS

National Curriculum and 5–14 Targets

England and Wales	Ma 2, 4e Ma 1, 4a
Northern Ireland	M 4d
Scotland	NMM Measure D2
Context	Groceries
within themes such as	The supermarket
Organisation	Group practically using resources, talking with classmates

Resources

- A variety of groceries which children can handle, for example, potatoes, fruit, vegetables, pasta shells, sweets.
- Plastic bags each labelled with an item and an amount, for example, 1 kg of apples, 250 g of pasta shells.
- Set of scales.
- A price list giving a price per metric unit for the groceries, for example:

   Potatoes – 90p per kilogram
   Apples – £1·80 per kilogram
   Bananas – £1·50 per kilogram
   Pasta shells – £1·30 per 500 g
   Sweets – £2·20 per 500 g

Key language

Estimate, kilogram, grams.

Aims

To provide experiences for children in estimating weights, weighing and calculating prices based on weight.

Objectives

Children should:
- estimate the amount for each specified weight
- weigh the item
- consult the price list
- calculate a possible price for each item.

Procedure

1 The group should be presented with about five carrier bags or boxes, each with an item such as mushrooms which can be estimated and weighed. They are to use a set of scales, plastic bags, and labels. The children are also given a price list.

2 Explain that they are to pretend they are working in a supermarket. Form three pairs within the group. Tell the first pair that their task is to estimate and make up the plastic bags with the amount of each item specified on the label, for example, 2 kg potatoes, 150 g of sweets. The second pair are given the set of scales and their task is to change each estimated bag to as accurate a weight as possible. The third pair are to put the cost on the label, using the prices on the list for their calculation. These bags should be set aside.

3 Suggest that the pairs may like to change tasks and make up a second set of bags.

4 The two bags of each item should be checked to find out if they are the same weight and that they are priced with the same amount. Changes should be made where necessary.

MORE POCKET MONEY

APPROXIMATE AGE GROUP
10-YEAR-OLDS

National Curriculum and 5–14 Targets

England and Wales	Ma 5, 4c
Northern Ireland	D4 b
Scotland	IH Interpret Ec
Context	Pocket money
within themes such as	Ourselves
Organisation	Class, or group, listening to the teacher

Resources

None required. (Some children may use calculators if this would be helpful for the addition when finding the mean value.)

Key language

Mode, median, mean.

Aims

To consolidate the language of mode and median as well as introduce the term mean.

Objectives

Children should:
- be able to explain the terms mode and median
- understand and explain the mean value.

Procedure

1 Explain to the children that William has a problem. He wants more pocket money from his Mum and she has told him he must prove to her that he is being given less than his classmates.

2 Ask the children what William should do. They are likely to respond that he should find out what the others get for pocket money.

3 Tell the pupils that William asked the first ten classmates he met the next morning and here is his record of their pocket money:

Pocket money of 10 pupils in my class		
Amount		*How many*
£2	////	4
£2·50	//	2
£3·50	//	2
£4·70	/	1
£5·50	/	1

4 Ask the children what they can find out about pocket money from this data. They might mention the mode or the median, terms which they have met before. If not you may have to remind them of each of them, for example: 'What does *mode* mean?'

5 The children should recall that the mode is the most frequent amount and find that for the pocket money, the most usual amount is £2. Explain that this makes William very gloomy as his pocket money is £2·50! He does not want to tell Mum about the mode amount for the group.

6 Encourage the children to recall the term *median* and suggest that William's Mum might be convinced if he has less than the middle person in the group. Ask them to find the median and have one child explain how he or she did this. The middle value is between the fifth and the sixth amounts:

£2, £2, £2, £2, **£2·50**, **£2·50**, £3·50, £3·50, £4·70, £5·50

You may need to remind some children that the median, when there is no middle amount, is found by adding the two middle amounts and halving the sum. Here the median pocket money is £2·50. Conclude that William is unlikely to get more pocket money if he uses the median amount!

7 Explain that a friend of William's told him that his Dad based his pocket money on the *mean* value. Let us find out about the mean value. To find the mean you add up all the amounts and then pretend to share the money out equally among everybody. This 'pretend equal share' is the mean. Ask the

children to find the total and then the mean value. Record each step for them:

Total pocket money = £30·20
Mean amount   = £30·20 ÷ 10
              = £3·02

8 Tell the children that William was delighted and went home to explain to Mum how he had carried out a survey and found the mean value. Ask someone to explain what the mean is and how it is found, as William will do. Ask the children if they think William will get more pocket money.

WHOLE NUMBERS – PLACE VALUE

APPROXIMATE AGE GROUP
10- OR 11-YEAR-OLDS

National Curriculum and 5–14 Targets

England and Wales	Ma 2, 4b
Northern Ireland	N4a
Scotland	NMM Range D1a

Context within themes such as	Large numbers Population, sport
Organisation	Class or group talking with the teacher

Resources

- Newspaper headlines.
- Base Ten number materials.

Key language

Unit, ten, hundred, thousand.

Aims

To investigate the number system for thousands and attempt to give some notion of magnitude of such amounts.

Objectives

Children should:
- know the place value system
- represent numbers with the Base Ten materials
- be able to read numbers up to 999999
- be able to relate large numbers to some references.

Procedure

1 Ask someone in the class to write how many children are in the class, for example, 29. Ask for an explanation of what the numerals mean, that is, two sets of ten and one set of nine. Check this by grouping the seated children in the two tens and the nine.

2 Ask a child to read a number headline, for example, 'Factory will have 2500 new jobs . . .' Write the number for the others to see. The

child can explain how to read the number. Establish that you look at the columns and use the value names, that is:

thousands	hundreds	tens	units
2	5	0	0

so we read 'two *thousand*, five *hundred*'. Discuss how we don't use the 's' in the value names and we don't need to mention the tens and units here because there aren't any.

3 Give the pupils a number such as 3043 and discuss how we *don't* say 'three thousand, four ten and three unit'. This links to the fact that we don't use the word unit(s) when we read numbers and we have 'short' versions of four tens, like forty.

4 Consider how we might represent two thousand five hundred and forty-three so that we could see that amount. Conclude that it would need to be something quite small for us to 'see' the number, possibly, dots or people's heads in a large crowd might be used. Remind the children, if they have not already mentioned it, that we have a representation of our number system using a small cube as one unit.

5 Look at the structured pieces and find the value of the long to be ten, the flat to be a

hundred and the block to be a thousand. Tell the children that some people think the block represents six hundred and ask them why they think that. Responses should mention that those are the cubes that can be seen on the six faces of the block and that the ones 'inside' had been forgotten. Ask a child to prove that there were a thousand cubes.

6 Ask for another headline to be read, for example, '106 000 doctors and dentists . . .' Discuss how the number system is extended to give us units and tens and hundreds of thousands like this:

thousands			ones		
hundreds	tens	units	hundreds	tens	units
1	0	6	0	0	0

We say this as 'one hundred and six thousand'. Give the children other examples to write and read.

7 Ask the children how we can represent numbers like these. Discuss how you would need larger pieces. Form a long-block with ten blocks to see what ten thousand would look like.

8 Ask the children to think of real life references for a thousand, ten thousand and one hundred thousand.

PERCENTAGES (1)

APPROXIMATE AGE GROUP
10- OR 11-YEAR-OLDS

National Curriculum and 5–14 Targets

England and Wales	Ma 2, 5b
Northern Ireland	N4f, 5c
Scotland	NMM Fractions D
Context	A sale
within themes such as	Shopping
Organisation	Group watching, listening and discussing with the teacher

This is the first of two lessons which a teacher could develop with a group.

Resources

• Sales adverts.

Key language

Percent, percentage, sale, reduced, reduction.

Aims

To introduce the concept of percentage.

Objectives

Children should:
• extend their vocabulary – sale, reduction
• understand that a percentage is a fraction of an amount
• know that percent means per hundred
• know that, for example, 1% is one hundredth of an amount.

Procedure

1 Show the children the adverts, giving them time to pick up one or two and read them.

2 Ask them to tell you what all the adverts have in common. Discuss their replies so that children extend or consolidate their understanding of a term like 'sale'.

3 Discuss different ways in which the shop tells people about lower prices, for example, giving the recent price crossed out with the reduced price beside it, by simply stating a price as a special offer, by stating all items in the shop are reduced by a number with a sign like this %.

4 Ask the children if they know the sign %. Discuss that it means the word 'percent' and that this means 'per hundred'. Establish that, for example, a 10% reduction is 'a bit off the price'. Explain to the children that the reduction could be given as a common fraction and a shop could say 'one tenth' off all the items. Discuss with the children why a different form of fraction called a percentage is used and establish that if a fraction with the same name is always used it makes it easier to compare the fractions. Remind them that it is difficult to compare, for example, a third and a half. Ask them if an advert stated 'one half off all our beds' and another 'one third off all our beds' would they know which is the better reduction? Some might have to think of one half as three sixths and one third as two sixths before being sure of the difference. This illustrates the need for fractions with the same name.

5 Explain that the name which has been chosen for comparing is hundredths. Ask the children to explain the fraction one hundredth and establish that it is one of one hundredth equal parts into which a whole has been divided.

6 Return to the adverts and ask for explanations for percentages such as 10%, and 40%. Establish that the price of an item is divided by a hundred to find one hundredth and the reduction is ten hundredths (ten times the one hundredth) or forty hundredths.

7 Calculate a 10% reduction on an item mentally, for example, 10% of £250 as

- one hundredth of £250 is £2.50 (divide by the digits moving two places to the right)
- ten hundredths are £25 (again multiply by digit movement, one place to the left).

8 Explain that you will be discussing percentages again and looking at different ways of calculating them. Ask them to bring adverts they see in papers and magazines where there is a percentage shown.

PERCENTAGES (2)

APPROXIMATE AGE GROUP
10- OR 11-YEAR-OLDS

National Curriculum and 5–14 Targets

England and Wales	Ma 2, 5b
Northern Ireland	N4f, 5c
Scotland	NMM Fractions D

Context	Using resources
within themes such as	Percentages

Organisation	Group practically using resources

Resources

- Squared paper and scissors.
- A packet of 100 envelopes and a packet of 500 sheets of paper.
- £1 in coins (real, card or plastic) in one bag and £1·50 in another.
- A packet of 100 paper clips and a packet with 300 drawing pins.
- Envelopes and slips of paper.

Key language

Percent, percentage.

Aims

To extend children's concept of percentage.

Objectives

Children should:
- know that percent means per hundred
- know that, for example, 30% is thirty hundredths of an amount.

Procedure

1 Set out each of these instruction cards with the appropriate materials around a table or at desks.

You need squared paper and scissors.

Cut out a rectangle made up of 100 squares.
Cut off 20% of the rectangle.

Cut out a rectangle of 200 squares.
Cut off 20% of the rectangle.

Write your names on an envelope. Put all your pieces of squared paper in it.

Write your names and answers on paper.

1 You need a packet of 100 envelopes. Find how many envelopes are 10% of 100 and write the answer.

2 You need the packet of 500 sheets of paper. How many sheets are 10% of 500?

Write your names and answers on paper.

1 You need the packet with 100 paper clips. Find how much is 25% of 100 and write the answer.

2 You need the packet with 300 drawing pins. How many drawing pins are 25% of 300?

Write your names and answers on paper.

1 You need the bag with £1. Find how much is 30% of £1 and write the answer.

2 You need the bag with £3. How much is in 30% of £3?

2 Discuss each activity and the children's answers with the whole group. Follow up any wrong responses by talking with the sub-group to identify the difficulty.

81   53